RESI
ENVIRONMENTAL CONTROL ENDORSEMENT

This book was set in Futura MD BT, Times New Roman, and Arial by Cathy J. Boulay, ETG.

Written by   Max Main

Graphics by   Michael R. Hall
Tristan J. Weetch

ISBN-13: 978-1-58122-105-3
ISBN-10: 1-58122-105-3

MARCRAFT

1 2 3 4 5 6 7 8 9 10   13 12 11 10 09

PREFACE

PURPOSE

Marcraft International has been producing certification courseware for different facets of the electronic, computer architecture, computer repair, and IT industry since 1988. This book has been designed to prepare students for the ETA RESI Environmental Control Certification exam, and provide them with the necessary tools for grasping the complex technologies of home automation.

The science of automated home technology has captured the imagination of consumers, standards groups, equipment suppliers, and home designers around the world. In only a few years, the home automation industry has grown rapidly from a hobbyist pastime to a mature technology that promises to provide increased efficiency in such areas as home energy management, heating and air conditioning management, home office networking, telecommunications, entertainment systems, irrigation systems, security systems, residential access systems, and more.

The home office has become a valuable and necessary asset for many homeowners. The introduction of wireless networking products and the "no new wires" network design standards have opened the door for new, low-cost design alternatives for the home computer user. Existing phone lines, or power line wiring, can now serve as the home network media supported by a wide array of industry standardized "plug and play" devices.

New Telecommunications Industry Association (TIA) standards and Federal Communications Commission (FCC) rulings issued in recent years have recognized the growing need for residential, as well as commercial, wiring standardization. The emergence of TIA-endorsed structured wiring designs for the whole home has rapidly become the main platform for integrating all of the home control and automation features into a cohesive home management system. The FCC has established a minimum grade for residential telecommunications cable installed in new homes. Until a few years ago, each residential automation or control technology required its own type of wiring and unique topology. Now, a single structured wiring technology standard has established a goal for supporting all the major new home automation technologies.

With the rapid growth of the Internet, faster desktop PCs, and the emergence of streaming audio and video, MP3, online game playing, e-mail, and improved residential broadband Internet access methodologies, the home office user can now utilize many of the advanced networking features previously available primarily for the corporate user.

It is now the task of the home technology integrators who are using this book to master the art of designing, installing, and programming the various home controllers, gateways, routers, home theater systems, and telecommunications devices that will rapidly become the core feature in the "smart home" of tomorrow.

RESI ENVIRONMENTAL CONTROL CERTIFICATION

The Electronics Technicians Association, International (ETA®) is an organization that establishes certification criteria for service technicians in the computer industry. With the infusion of IT into consumer electronic products, digital systems have become more popular and more powerful than ever before.

Please visit *www.etainternational.org* to view the ETA exam competencies and practice exams for environmental control.

Examinees have two ways to sit for a certification exam with ETA International. They may take an exam using either a paper or online format under the supervision of an ETA approved exam proctor. Exam results are returned to examinees within ten business days when paper exams are received at ETA's home office. Immediate results are displayed when the examinee completes the online exam.

ETA International currently has 850 ETA Certification exam proctors across the US and worldwide. To take an exam, locate an exam administrator (CA) and exam fees, please visit *www.etainternational.org* or call 1-800-288-3824 for more information. If your educational facility is interested in proctoring ETA exam, please call 1-800-288-3824 or e-mail us at *eta@eta-i.org*.

Key Features

The pedagogical features of this book were carefully developed to provide readers with key content information, as well as review and testing opportunities. A complete RESI Environmental Control objectives map and various Test Tip information boxes are included to help students key in on RESI specific materials.

RESI Environmental Control Exam Coverage

The locations of RESI Environmental Control specific materials are identified with a Test Tip marker in the margin that helps students focus on key content that they will be expected to know for the RESI Environmental Control Certification exam. Appendix A provides a comprehensive listing of the RESI Environmental Control objectives.

Pedagogical Features

A wealth of figures, diagrams, and screen dumps are included to provide constant visual reinforcement of the concepts being discussed.

The book begins with a list of learning objectives that establishes a foundation and systematic preview of the book. In addition, each book concludes with a book summary and a key-point review of its material.

Key terms are presented in bold type throughout the text. A comprehensive glossary of terms provides quick, easy access to key term definitions that appear in each book.

In general, it is not necessary to move through this text in the same order that it is presented. Also, it is not necessary to teach any specific portion of the material to its extreme. Instead, the material can be adjusted to fit the length of your course. As a matter of fact, practicing IT professionals can use the material in the CD test banks to identify the areas where they need to brush up, and then use the text book to study that material directly.

Teacher Support

An instructor's guide accompanies the course. Answers for the end-of-book quiz questions are included along with a reference point in the book where a particular item is covered. Sample schedules are included as guidelines for possible course implementations. Answers to all lab review questions and fill-in-the-blank steps are provided so that there is an indication of what the expected outcomes should be.

An electronic copy of the textbook is included on the Instructor's Guide CD-ROM disc.

Test Taking Tips

The RESI Environmental Control Certification exam is an objective-based timed test. It covers the objectives listed in Appendix A in a multiple-choice format. There are two general methods of preparing for the test. If you are an experienced technician using this material to obtain certification, use the testing features at the end of each book and on the accompanying CD to test each area of knowledge. Track your weak areas and spend the most time concentrating on them.

After completing the study materials use the various testing functions available on the CD to practice taking the test. Use the Study and Exam modes to test yourself by chapter, or on a mixture of questions from all areas of the text. Practice until you are very certain that you are ready. The CD will allow you to immediately refer to the area of the text that covers material that you might miss.

- Answer the questions you know first. You can always go back and work on previous questions.

- Don't leave any questions unanswered. They will be counted as incorrect.

- There are no trick questions. The most correct answer is in there somewhere.

- Be aware of questions that have more than one correct answer. Questions that have multiple correct answers can be identified by the special formatting applied to the letters of their possible answers. They are enclosed in a square box. When you encounter these questions, make sure to mark every answer that applies.

- Get plenty of hands-on practice before the test, using the time limit set for the test.

- Make certain to prepare for each test category listed above. The key is not to memorize, but to understand the topics.

- Take your watch. The RESI Environmental Control Certification exam is a timed test. You will need to keep an eye on the time to make sure that you are getting to the items that you are most sure of.

- Get plenty of rest before taking the test.

All ETA exams are closed-book exams, which means that you will not be permitted to take any material into the testing area. However, you will be provided with a blank sheet of paper and a pen or pencil. When you complete a ETA certification exam, the test center software informs you whether you have passed or failed the exam.

Table of Contents

LAB PROCEDURES

LAB PROCEDURE 3 – AN AUTOMATED MULTIZONE X-10 POWER AND CONTROL SYSTEM

LAB PROCEDURE 4 – RESIDENTIAL POWER AND CONTROL SYSTEMS RESEARCH

LAB PROCEDURE 5 – WHOLE-HOUSE LIGHTING CONTROL AND MANAGEMENT SYSTEMS

LAB PROCEDURE 6 – STRUCTURED WIRING

ENVIRONMENTAL CONTROL

Upon completion of this book and its related lab procedures, you should be able to accomplish the following tasks.

- Identify user interfaces and their appropriate applications.
  - Device types
  - Describe the importance of simplicity and ease of use as it pertains to the end user.

- Define and recognize control systems which integrate subsystems in the home.
  - Describe their functionality, characteristics and purpose.
  - Embedded control systems and Personal computer (PC) based control systems

- Identify commonly used communication protocols and their application.
  - IR
  - Serial
  - IP
  - RF
  - Bluetooth
  - Contact closure
  - Inputs (zones)
  - Z-wave and Zigbee
  - ASCII
  - Proprietary protocols

- Describe basic HVAC (Heating Ventilation and Air Conditioning) terminology and install peripheral control devices.
  - Control layer
  - Communication layer
  - Zones HVAC
  - Programmable thermostats
  - Importance of referencing manufacturer specification and compatibility.

- Describe basic lighting terminology and install peripheral control devices.
  - Identify lighting control applications
  - Communication interface/bridge
  - Identify lighting control protocols (Open standards)
  - Proprietary RF and proprietary low voltage

- Identify and install component power protection devices.
  - Identify whole house protection options
  - Identify and install point protection

Environmental Control

This book will help you become familiar with the protocols, devices and applications that are used for home control and management. Topics include input/output devices used for HVAC and home security systems, telephone systems and home lighting management systems.

RESIDENTIAL CONTROL SYSTEMS

Not so long ago, television manufactures sold television sets with remote controls as an option. Today you cannot buy a television that does not include a remote controller. The continuing movement in all areas of residential consumer electronics is to provide more sophisticated control systems to make control of the various residential systems more convenient for the user.

At the low end of this movement, simple controllers are used for each A/V device, the security system, the lighting system and the environmental controls. A step up from this involves programming a **learning remote controller** with information to control multiple devices and systems from a single source. At the top of the movement, all of the residential subsystems can be controlled through a centralized whole-home control system. Home control systems provide the capability to manage audio/video systems. lighting, HVAC and security systems. A single device such as a desktop or laptop computer may be used to control all of the home control systems from a single control source, Web pads, PDAs and smart phones are also appearing on the home technology market for controlling individual home automation systems.

While the controller is centralized, the control element is not. These systems provide the user with the ability to control the residential subsystems from anywhere within the home, or even from remote locations outside the home. In some instances, this control can be effected from anywhere in the world using the telephone system or the Internet.

Residential Controllers

As shown in Figure 1-1, a **controller** integrates all of the automated home features into a coherent system. Residential controllers are more elaborate than gateways and are designed to manage multiple whole-home systems such as irrigation, lighting, HVAC, security, and audio/video from a central location. Commands are entered by the user from a wall-mounted keypad, a handheld infrared remote keypad, or a PC. Output status display information and system menus from the controller are provided to a PC or TV.

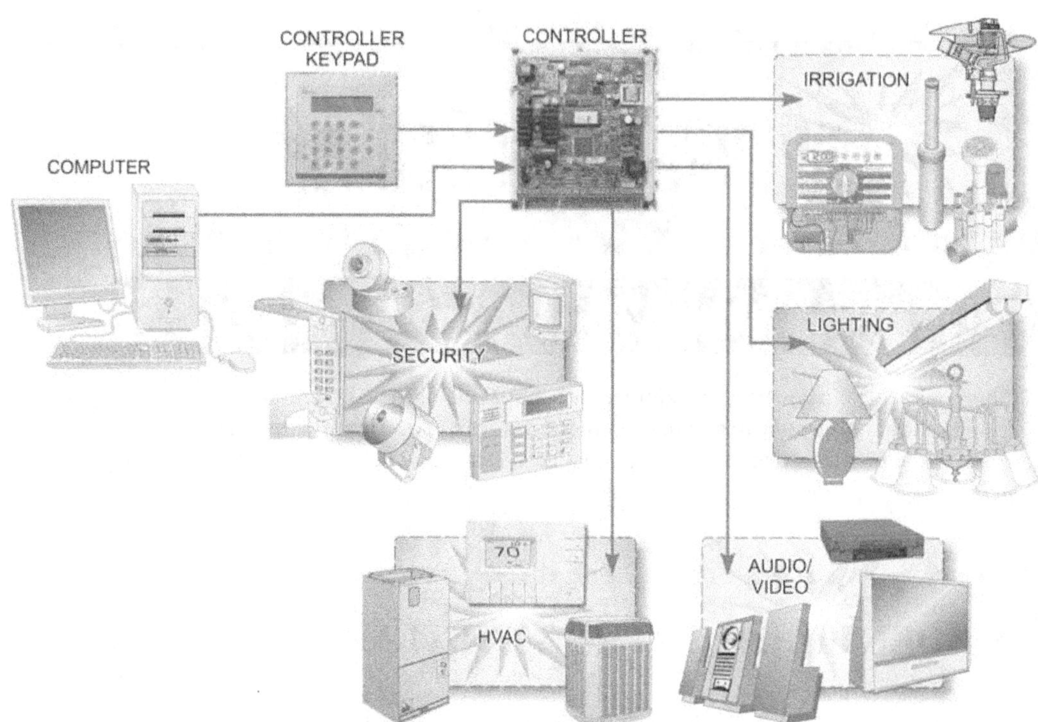

**Figure 1-1:
Home Controller
Configuration**

Home controller motherboards, input/output jacks, connectors, and power supplies are usually mounted in a control cabinet as shown in Figure 1-2.

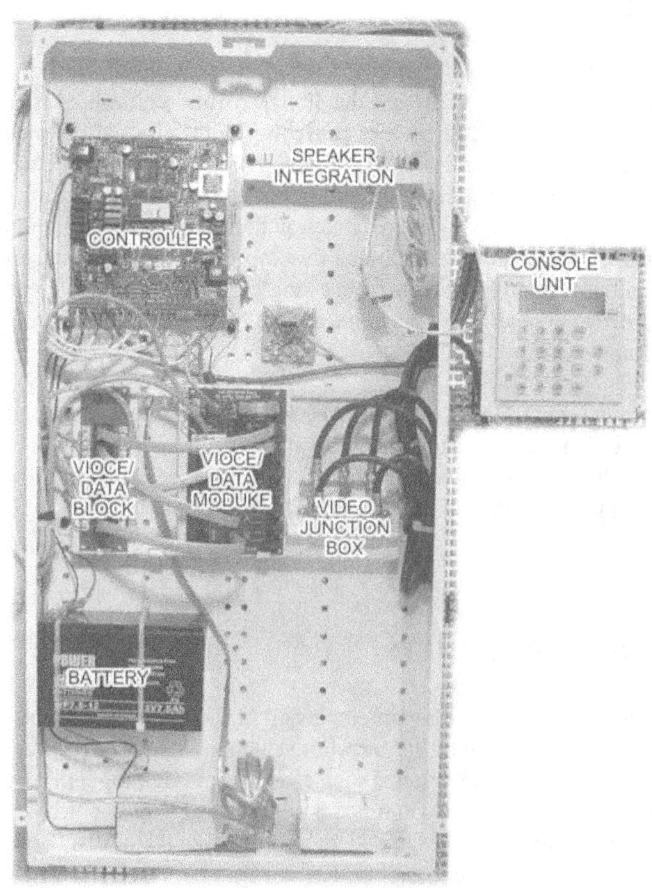

**Figure 1-2: Home
Controller Cabinet**

Typical input and output interface ports supported are:

- F-type coaxial connectors for audio/video distribution

- RJ-45 UTP ports for interconnection with the home office computer network

- A serial data port (RS-232 or RS-485) for remote location communications

- A interface to the home power line system (typically X-10 controller modules)

- A port for a backup processor unit

- An expansion port that supports off-the-shelf Type II compact flash memory

Two Electronic Industries Alliance (EIA) standards are frequently included as a built-in feature with home controllers for interfacing with external devices. They are the CEBus and EIA RS-485 serial bus interface specifications. CEBus (Consumer Electronics Bus) is an open standard for home automation released by the EIA as EIA-600.

The structured wiring cabinet may also hold the residential network's router and digital modem. The presence of these devices in the structured wiring cabinet makes it easy to connect and access the residential controller through the residential network or the Internet. This opens up the residential controller to control from outside the residence — such as through the Internet. Some residential controllers possess networking connectivity that enables them to connect directly to the Internet through the router and modem. Other versions may require the presence of an always-on host computer to provide the interface between the Internet and the controller.

Keypads

The **keypad** is a device equipped with a numerical set of pushbuttons that is similar to a telephone touchtone keypad. Keypads are used for programming, controlling, and operating various residential access control and management devices. They can be located in any area of a home that is convenient for the resident to operate external gates and doors. Home residents can use keypads to initiate commands for control options such as arming and disarming the system or bypassing a zone. A controller keypad is shown in Figure 1-3.

Figure 1-3:
Controller
Keypad

Figure 1-4: A
Remote Control

Remote Controls

Remote controls are used to issue commands from a distance to televisions or other consumer electronics such as stereo systems and DVD players. Remote controls for these devices are usually small wireless handheld objects with an array of buttons for adjusting various settings such as television channel, track number, and volume, as shown in Figure 1-4. However, remote controls can also be quite complex incorporating programmable specialty buttons and icon-driven touchscreens, as depicted in Figure 1-5.

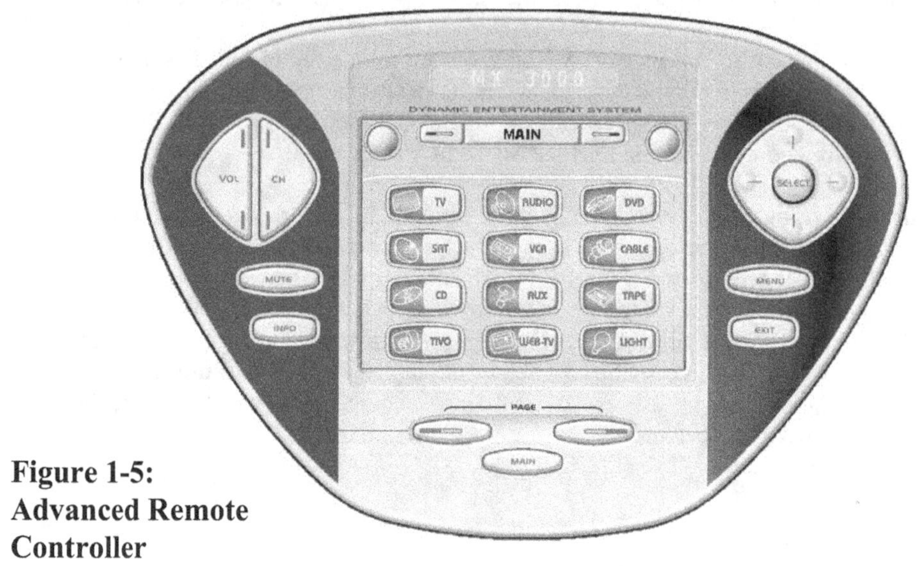

Figure 1-5:
Advanced Remote
Controller

Touchscreens

A **touchscreen** enables users to operate home automated systems by simply touching icons on a display screen. A touchscreen remote controller is shown in Figure 1-6.

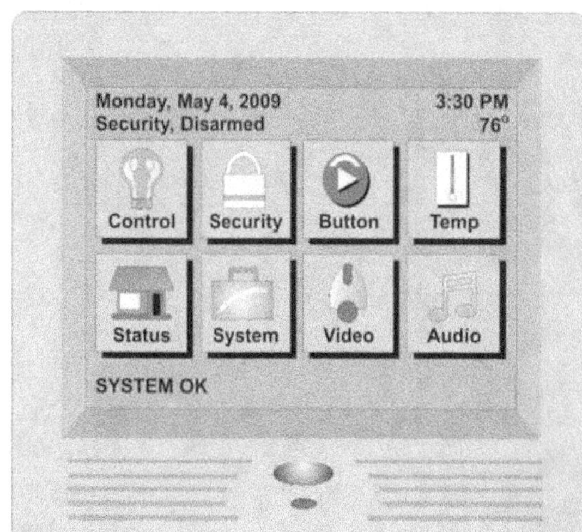

Figure 1-6:
Touchscreen
Remote Controller

The touchscreen interface has been widely available for several years and recently has been adopted by the home networking industry. It enables users to navigate a home network management system by touching icons or links on the screen. Touchscreen controllers can be programmed to meet the requirements of most custom integrated home systems such as audio and video systems, security systems, and home light management systems.

Some newer home controller systems include a color touchscreen interface. This provides a consumer-friendly solution that's simple to operate. The size of a standard screen varies from 8 to 10 inches. These systems are available in panel format and can be flush-mounted in a wall or installed in attractive real-wood tabletop valets.

More advanced touchscreen controllers include a high-quality color monitor that enables you to view video sources such as cable TV, satellite TV, and closed circuit TV cameras. Although they are expensive, when used in the right locations, they avoid the need for several multibutton switches and keypads.

In a distributed management configuration, touchscreen interfaces can be located wherever the homeowner needs system control capabilities.

Keyfobs

A **key fob**, as shown in Figure 1-7, is an item people often carry with their keys, on a ring or a chain. Key fobs are often called "key rings" or "key chains" in colloquial usage. Electronic key fobs are used for remote keyless entry systems on motor vehicles. Early electric key fobs operated using infrared and required a clear line of sight to function. These could be copied using a programmable remote control. More recent models use challenge-response authentication over radio frequency, so these are harder to copy and do not need line of sight to operate. Programming these remotes sometimes require the automotive dealership to connect a diagnostic tool but many of them can be self-programmed by following a sequence of steps in the vehicle and usually requires at least 1 working key.

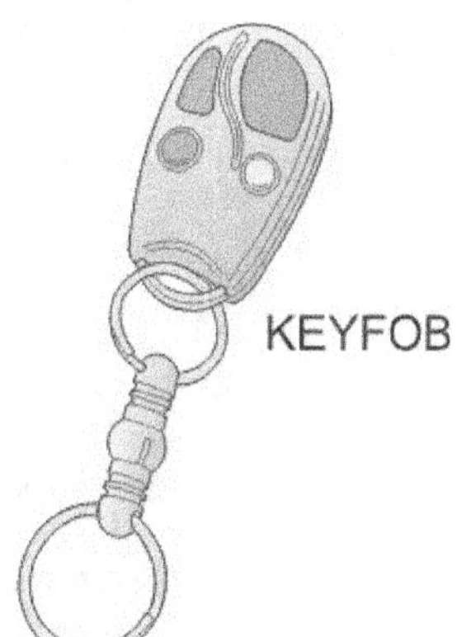

Figure 1-7: Keyfob

Tablet Computers and Webpads

A tablet computer is a complete computer contained in a touchscreen. **Tablet PCs** can be specialized for only Internet use or be full-blown, general-purpose PCs. The distinguishing characteristic is the use of the screen as an input device using a stylus or finger. A tablet PC is shown in Figure 1-8.

Figure 1-8: A Tablet PC

A **webpad** is a handheld, wireless device designed for Web browsing. A Webpad is like a laptop computer without a fold-down screen and keyboard, which is why it is also called a "Web slate" or "tablet computer." Using a touchscreen, webpads typically weigh under three pounds and are less than an inch thick. An example of a Web pad design is shown in Figure 1-9.

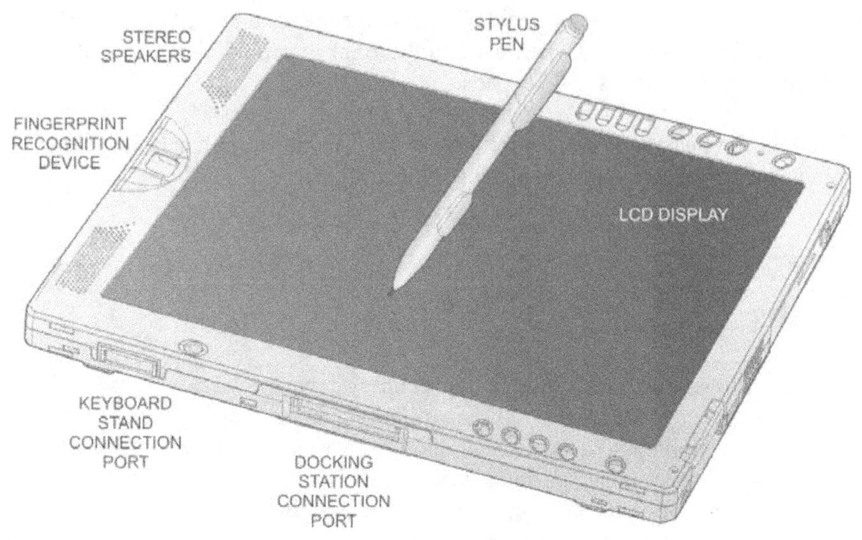

**Figure 1-9:
A Webpad**

Cell Phones and Smartphones

Unlike conventional wire-based cordless phones used in a residential application, **cellular telephones** are completely portable and do not require proximity to a jack to access the wire-based networks operated by local telephone companies. Cellular phone networks exist in nearly every metropolitan area throughout the world.

A **smartphone** is a mobile phone offering advanced capabilities beyond a typical mobile phone, often with PC-like functionality. The term smartphone is sometimes used to characterize a wireless telephone set with special computer-enabled features not previously associated with telephones.

In addition to functioning as an ordinary telephone, a smartphone's features often include:

- Wireless e-mail, Internet, Web browsing, and fax

- Intercom function

- Personal information management

- Online banking

- LAN connectivity

- Local data transfer between phone set and computers

- Remote control of computers

- Remote control of home or business electronic systems

- Interactivity with unified messaging

- Video conferencing

Figure 1-10. depicts services typically provided by smartphones.

**Figure 1-10:
Cell Phone Services**

PDA's

A **Personal Digital Assistant (PDA)** is handheld computer for managing contacts, appointments and tasks. It typically includes a name and address database, calendar, to-do list and note taker, which are the functions in a personal information manager (see PIM). Wireless PDAs may also offer e-mail, Web browsing and cellular phone service. Data are synchronized between the PDA and desktop computer via a cabled connection or wireless. A typical PDA is shown in Figure 1-11.

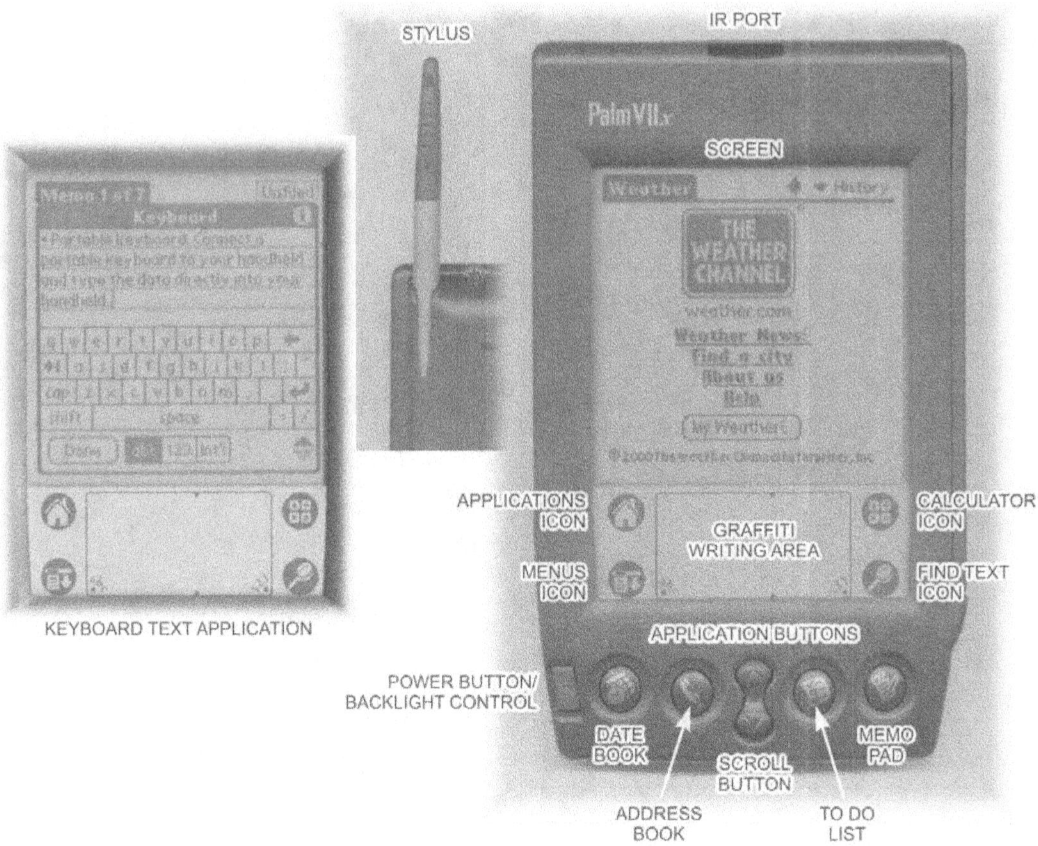

Figure 1-11: A Personal Digital Assistant

Networked Game Devices

With the advent of new wireless media receiver products, the networking of game devices together (**entertainment networking**) has become a powerful market. The development of driver-free game network adapters has helped to develop a new generation of game players (partners) in a way that transcends the idea of mere entertainment. Online games have connecting various international communities in a way that actualizes a previously virtual world. Figure 1-12 illustrates the integration of a typical game console into a residential network environment.

Wireless companies have come calling on video-game developers with the idea of combining the portability of cell phones with the power of virtual worlds to create a more intense gaming culture. In fact, a minimum requirement for every new cell phone now shipping is that it includes the capability of running networked gaming. This ensures the emergence of many new handheld devices that combine mobile gaming and cell phone connectivity. Clients will soon be playing full-color, three-dimensional, multiplayer games over cell phone networks.

A laptop console optimized for gaming, means that for game playing enthusiasts, the next game is no further away than a hip pocket. Products such as the game deck are cleverly marketed through multi-day competitions where competitors speaking many languages across dozens of countries square off playing various games. Players receive points for their speed and dexterity through time trials and problem solving. Rewards are often tied to players' abilities in connecting with each other and employing multi-player team strategies.

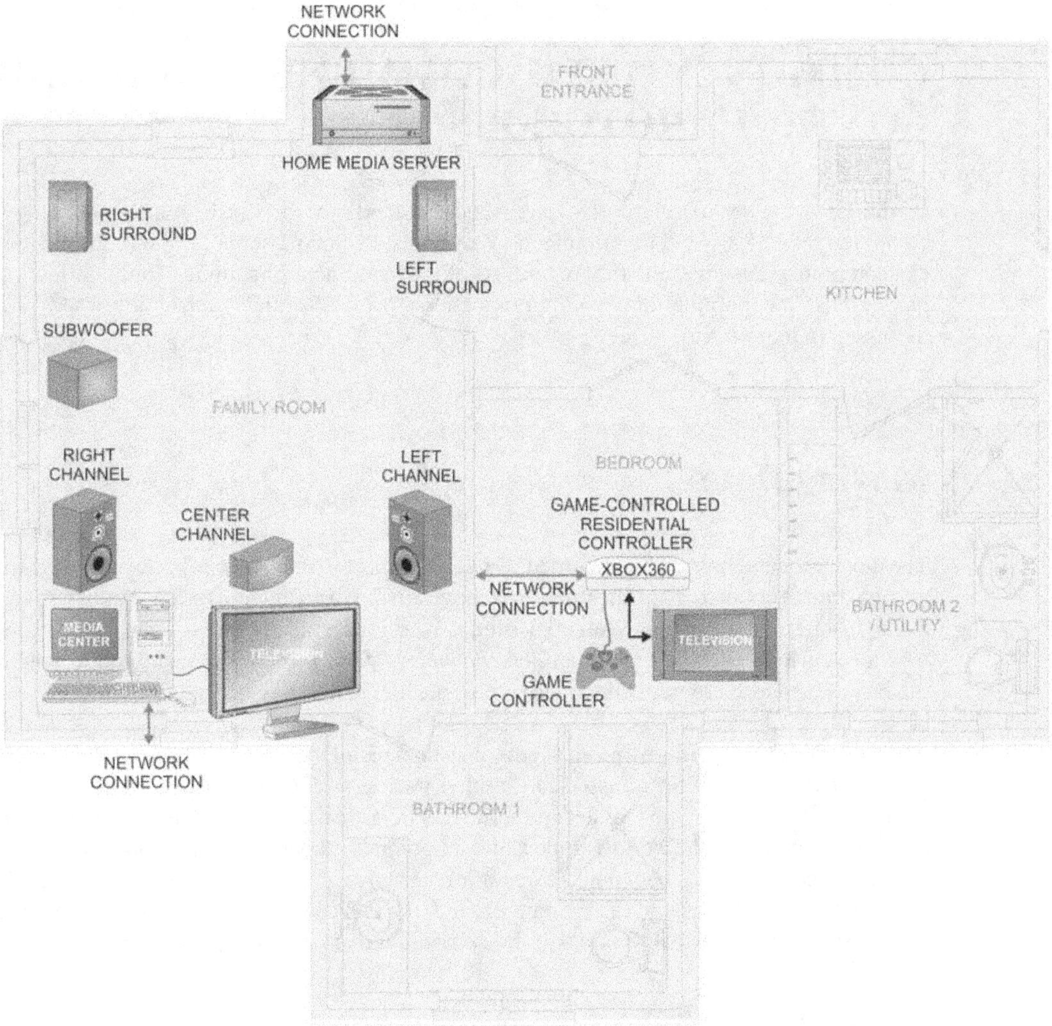

Figure 1-12: Game Controllers

Personal computers (PCs)

Personal Computers

Personal computers (PCs) have become an essential component of an automated home management system. Applications for home users include resource sharing, home automation and control systems, streaming audio and video entertainment, Internet telephony, digital photography editing, games, and e-mail. A typical PC is shown in Figure 1-13.

Wireless networks are a good choice for portable PC networking. Portable computers may be used throughout the home as long as they are within range of other wireless terminals or the wireless access point. Installing the network interface cards and network operating system software is all that is necessary to start sharing data and home network resources.

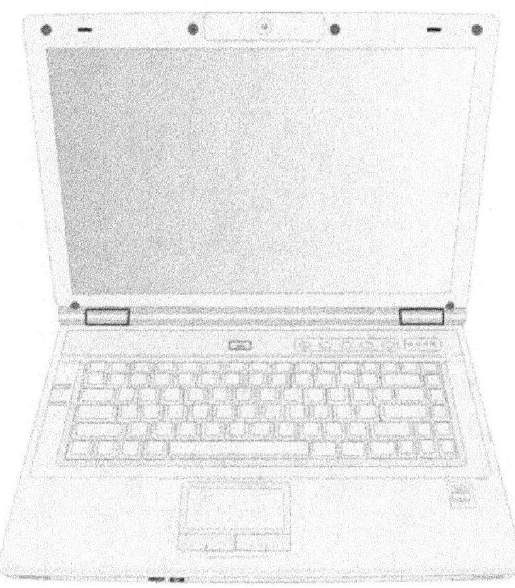

Figure 1-13: A Typical Portable PC

COMMONLY USED COMMUNICATION AND CONTROL PROTOCOLS

Protocols are a set of rules often published as standards for use in home technology, computer networking and communications systems. Protocols serve to assure all devices can communicate with each other regardless of the manufacturer/supplier. In the following paragraphs, the popular home communications protocols and control interfaces are discussed in more detail

Moving Data

The most frequent operation performed in a computer is the movement of information from one location to another. This information is moved in the form of words. These transfers can occur in one of two modes—**parallel mode**, where an entire word is transferred from location A to location B by a set of parallel conductors at one instant, and **serial mode**, where the bits of the word are transmitted along a single conductor, one bit at a time.

Serial transfers require more time to accomplish than parallel transfers because a clock cycle must be used for each bit transferred. Parallel transfers require only one clock pulse to complete. Examples of both parallel and serial transfers are depicted in Figure 1-14. Because speed is normally of the utmost importance in computer operations, all data movements within the computer are conducted in parallel, as shown in (a). But when information is being transferred between the computer and its peripherals (or another computer), conditions may dictate that the transfer be carried out in serial mode, as shown in (b).

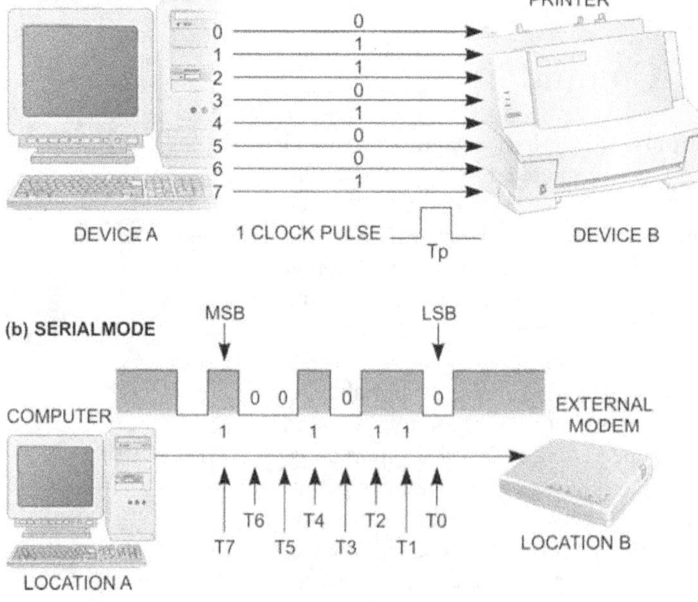

Figure 1-14: Parallel and Serial Data Transfers

Serial Interfaces and Protocols

Most serial ports on personal computers are **universal serial bus (USB)** or **IEEE-1394 FireWire** ports. However, there are still many residential devices that employ older serial interface types including the RS-232C, RS-422, and RS-485 standards.

Universal Serial Bus

The most widely used I/O interface scheme in current PCs is the universal serial bus (USB). This high-speed serial interface has been developed to provide a fast, flexible method of attaching up to 127 peripheral devices to the computer. USB peripherals can be daisy-chained or networked together using connection hubs that enable the bus to branch out through additional port connections. This connection format has been designed to replace the system's traditional serial- and parallel-port connections. A practical USB desktop connection scheme is presented in Figure 1-15.

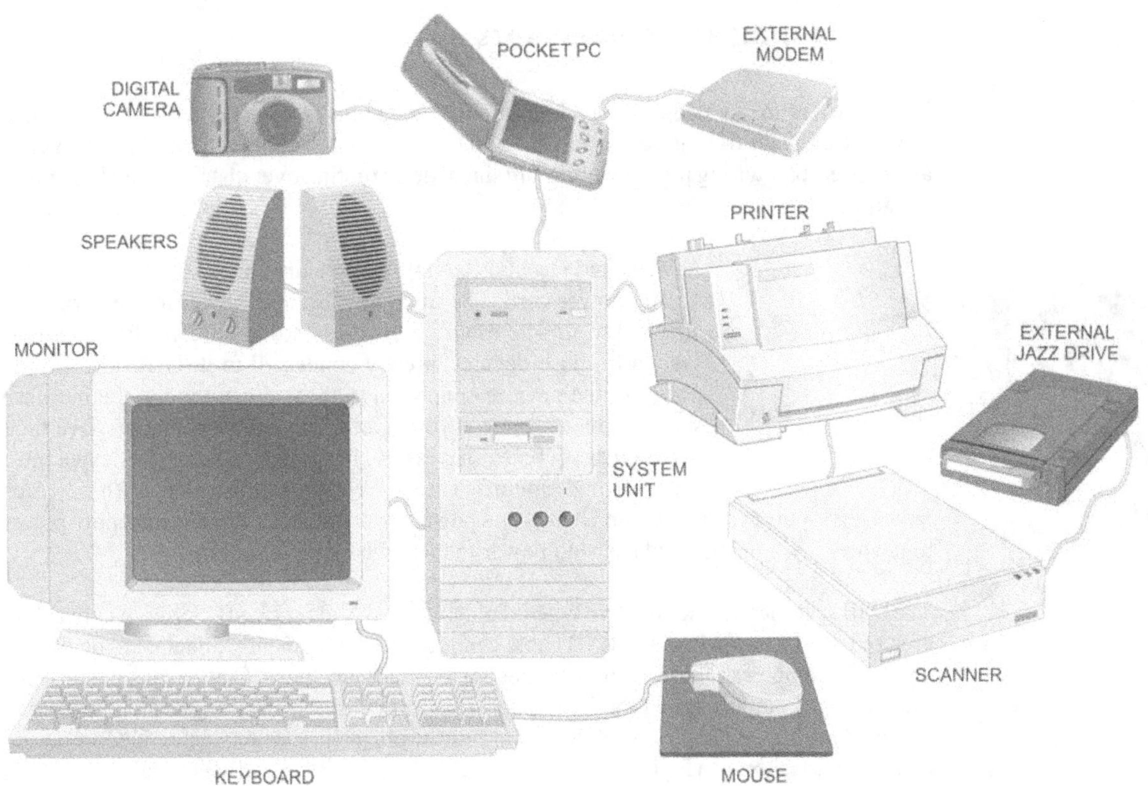

Figure 1-15: USB Connection Scheme

In this example, some of the peripheral devices are simply devices, whereas others serve as both devices and connection hubs. The system provides a USB host connection that serves as the main USB connection.

USB devices can be added to or removed from the system while it is powered up and fully operational. This is referred to as hot-swapping or hot-plugging the device. The Plug-and-Play capabilities of the system will detect the presence (or absence) of the device and configure it for operation.

The USB management software dynamically tracks what devices are attached to the bus and where they are. This process of identifying and numbering bus devices is known as **bus enumerating**. The USB specification promotes *hot-swap* peripheral connections that do not require the system to be shut down. The system automatically detects peripherals and configures the proper driver. Instead of simply detecting and charting devices at startup in a PnP style, the USB controller continuously monitors the bus and updates the list whenever a device is added to or removed from it.

While hot swapping USB devices typically does no harm to the system hardware or the device, the Windows operating system requires that you stop the device's operation using the Safely Removal Hardware tool. An icon for this tool appears in the notification area of the Windows task bar while the USB device is active. Failure to stop the device's operation before physically removing it from the USB port will cause Windows to generate an "Unsafe Removal of Device" error. It may also cause the system to crash and lose any unsaved data.

USB Cabling and Connectors

USB transfers are conducted over a four-wire cable, as illustrated in Figure 1-16. The signal travels over a pair of twisted wires (D+ and D–) in a 90-ohm cable. The differential signal and twisted-pair wiring provide minimum signal deterioration over distances and high **noise immunity**.

Figure 1-16: The USB Cable

A Vbus and Ground (GND) wire are also present. The Vbus is the +5Vdc power cord. The interface provides power to the peripheral attached to it. The root hub provides power directly from the host system to those devices directly connected to it. Hubs also supply power to the devices connected to them. Even though the interface supplies power to the USB devices, they are permitted to have their own power sources if necessary. In these instances, the device must be designed specifically to avoid interference with the bus power-distribution scheme. The USB host's power-management software can apply power to devices when needed and suspend power to them when not required.

The USB specification defines two types of plugs: series-A and series-B. Series-A connectors are used for devices where the USB cable connection is permanently attached to devices at one end. Examples of these devices are keyboards, mice, and hubs. Conversely, the series-B plugs and jacks are designed for devices that require detachable cabling (printers, scanners, and modems, for example). Both are four-contact plugs and sockets embedded in plastic connectors, as shown in Figure 1-17. The sockets can be implemented in vertical, right-angle, and panel-mount variations. Smaller 5-pin USB plugs and jacks (referred to as Mini-a and Mini-B) have been developed for the USB 2.0 specification, and an even smaller Micro-USB connector version has been proposed for the future. These connectors are intended for use with smaller devices such as digital cameras, cell phones and Personal Digital Assistants (PDAs). These connector selections are designed to provide rugged connections that are not prone to damage from repeated or incorrect usage.

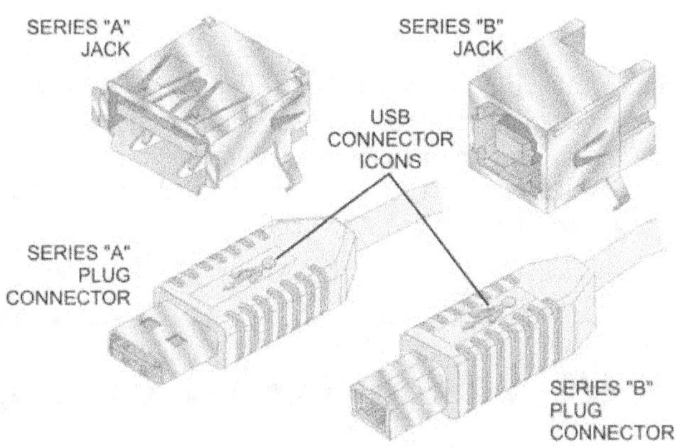

**Figure 1-17:
USB Connectors**

The connectors for both series are keyed so that they cannot be plugged in backward. All hubs and functions possess a single, permanently attached cable with a series-B connector at its end. The connectors are designed so that the A- and B-series connections cannot be interchanged.

IEEE-1394 FireWire Bus

While the USB specification was being refined for the computer industry, a similar serial interface bus was being developed for the consumer products market. Apple Computers and Texas Instruments worked together with the **Institute of Electrical and Electronic Engineers (IEEE)** to produce the **FireWire** (or **IEEE-1394**) specification. The bus offers a very fast option for connecting consumer electronics devices, such as camcorders, digital music players, and the original iPODs from Apple, to the computer system.

The FireWire bus is similar to USB in that devices can be daisy-chained to the computer using a single connector and host adapter. The entire interface requires a single IRQ channel, an I/O address range, and a single DMA channel to operate. FireWire is also capable of using the high-speed Isochronous transfer mode described for USB. The original IEEE-1394a FireWire specification (now referred to as FireWire 400) provides data transfer rates up to 400Mbps. However, an improved IEEE-1394b standard provides an additional electrical signaling method that permits data transmission speeds of 800Mbps (FireWire 800) and greater. Its high-speed capabilities make FireWire well suited for handling components, such as video and audio devices, which require real-time, high-speed data transfer rates.

A single IEEE-1394 connection can be used to connect up to 63 devices to a single port. However, up to 1023 FireWire buses can be interconnected. PCs most commonly employ a PCI expansion card to provide the FireWire interface. While AV equipment typically employs 4-pin 1394 connectors, computers normally use a 6-pin connector, with a 4-pin to 6-pin converter. The maximum segment length for an IEEE-1394 connection is 4.5 m (14 ft). Figure 1-18 depicts the FireWire jack and plug most commonly used with PCs.

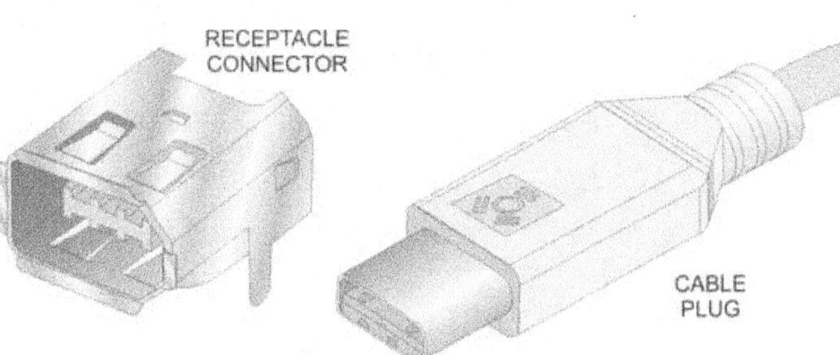

Figure 1-18: FireWire Plug and Jack

The IEEE-1394 cable is composed of two twisted-pair conductors similar to those used in the local area networks. The IEEE-1394b version of the standard also supports new transport media, including glass and plastic optical fiber, as well as Category 5 copper cable. With the new media comes extended distances; for example, 100 meters over CAT5 cabling. Like USB, FireWire supports both PnP and hot swapping of components. It also provides power to the peripheral devices through one pair of the twisted conductors in its interface cable.

A proposed version of the IEEE-1394 standard (titled *P1394b*) provides an additional electrical signaling method that permits data transmission speeds of 800 Mbps and greater. The new version of the standard also supports new transport media including glass and plastic optical fiber, as well as Category 5 copper cable. With the new media comes extended distances, for example, 100 meters over CAT5 cabling.

RS-232, RS-422, and RS-485 Serial Port Standards

The RS-232 standard has been renamed by the Electronic Industries Association (EIA). EIA has published three modifications, the most recent being the EIA 232F standard introduced in 1997. Besides changing the name from RS-232 to **EIA 232**, some signal lines were renamed and various new ones were defined, including a shield conductor. RS-422 interfaces are typically used when the data rate or distance criteria cannot be met with RS-232. Figure 1-19 depicts a typical 9-pin to 25-pin connection scheme. Notice the crossover wiring technique employed for the TXD/RXD lines displayed in this example.

The RS-422 standard allows for operation of up to 10 receivers from a single transmitter. RS-422 is rated to 10 Mbps and supports reliable communication at distances up to 4,000 ft. This standard also defines electrical characteristics that will allow one transmitter and up to 32 receivers on the line at once. The 15 m limitation for cable length can be stretched to about 30 m for ordinary cable, if well screened and grounded, and about 100 m if the cable is low capacitance as well.

The third RS (EIA) standard used in home automation and home theater environments is RS-485. RS-485's main features include:

- Differential (balanced) signaling with a minimum difference of potential of only 0.2V required for valid operation.

- RS-485 uses one pair of wires. The pairs may be shielded to protect against electromagnetic interference, but if shielding is not required, ordinary phone line can be used to carry RS-485 signals.

- The speed of RS-485 is reduced with cable distance, but RS-485 transmits at up to 10Mbps at 12 meters, and up to 100Kbps at its maximum distance of 1,200 meters (about 4,000 feet).

- RS-485 supports up to 32 drivers and 32 receivers per network. An RS-485 network uses a half-duplex (send or receive data) daisy-chained configuration.

- RS-485 can be referred to as the OSI Model Physical layer two-wire, half-duplex, multipoint serial connection.

RS-485 is often used as the connection type for residential controllers and PCs to connect intelligent control devices such as thermostats, keypads, and access control devices. RS-485 uses a point-to-point bus topology.

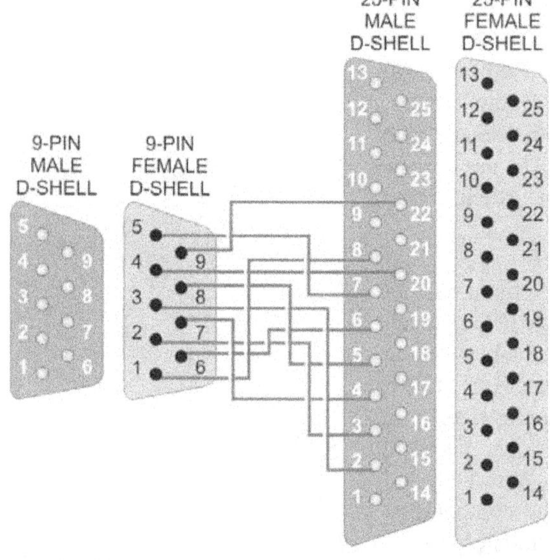

3	TX DATA	2
2	RX DATA	3
7	RTS	4
8	CTS	5
6	DSR	6
5	SIG GND	7
1	CXR	8
4	DTR	20
9	RI	22

Figure 1-19: A 9-Pin to 25-Pin RS-232 Cable

Figure 1-20 shows how RS-232, RS-422, and RS-485 can be used in a typical home automation configuration.

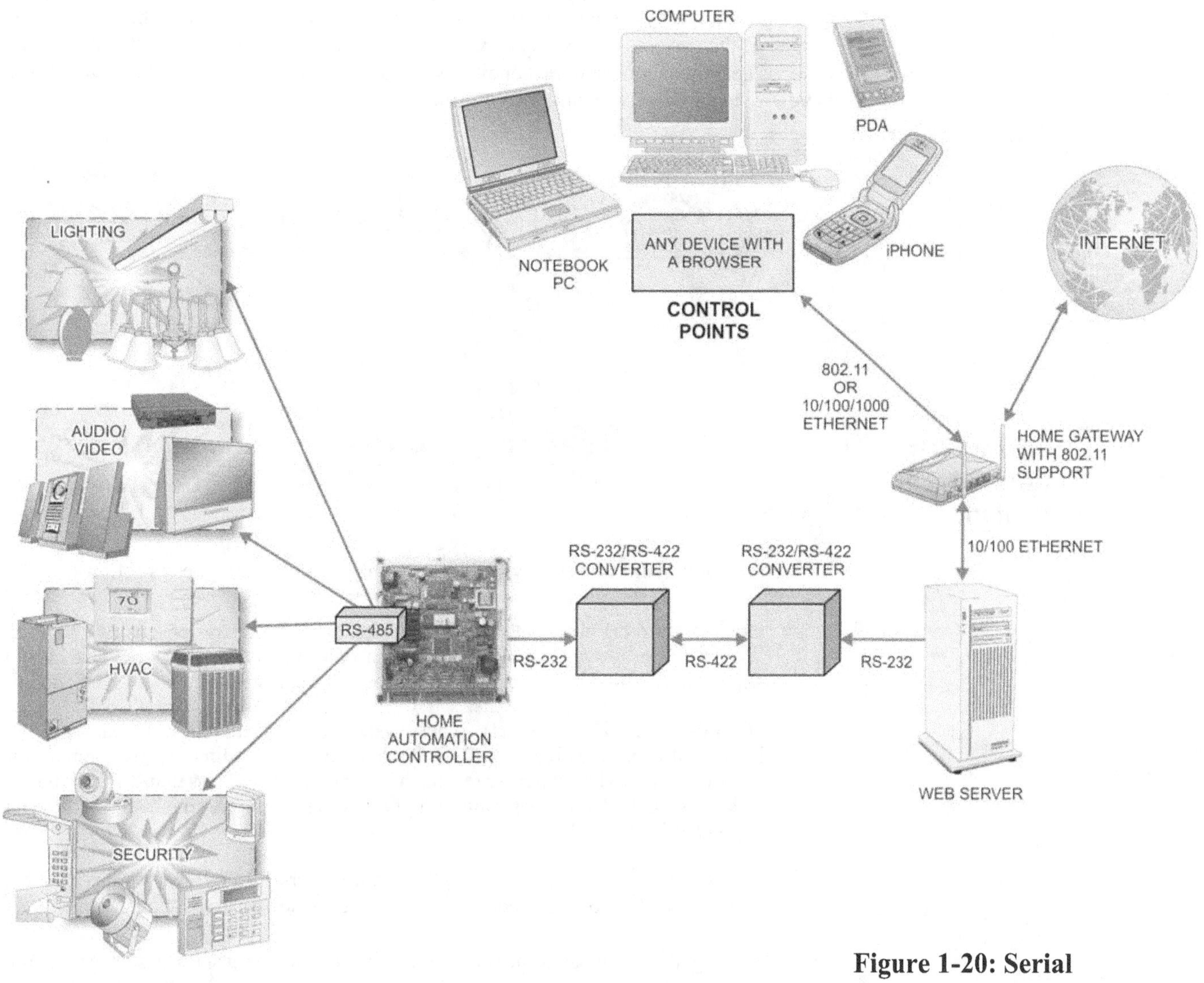

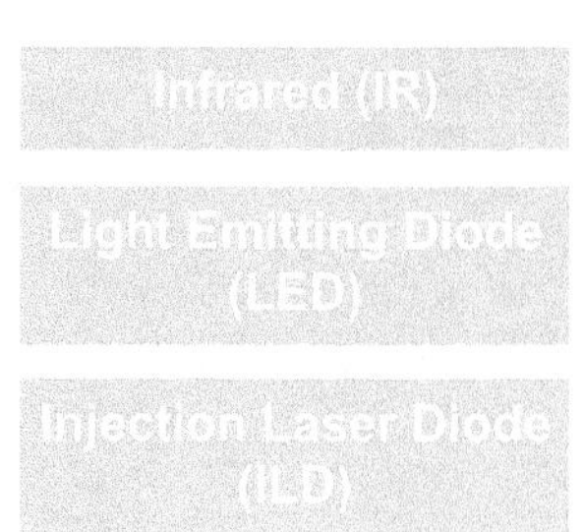

Figure 1-20: Serial Interface Applications

Infrared Protocols

Infrared (**IR**) systems carry data between devices using infrared light. The light source can be a **Light Emitting Diode** (**LED**) or an **Injection Laser Diode** (**ILD**). Photodiodes are required to receive the signals.

Infrared (IR)

Light Emitting Diode (LED)

Injection Laser Diode (ILD)

Most consumer AV equipment uses infrared light pulses to transmit instructions. A small eye called a capture is built into a unit that sense a type of light called infrared (IR) radiation. Humans can't see this light, but this eye can. When a control device sends a pre-adjusted pulse of IR, the eye senses this light, and based on the pulse code, interprets what operation it is supposed to carry out.

Infrared light uses the THz (1,000 GHz) range of frequencies, and therefore provides high bandwidths of up to 20Mbps. Figure 1-21 shows the frequency range associated with infrared light. Unfortunately, these high frequency waves provide no penetration ability and are readily diluted by sunlight. The cost of infrared systems varies considerably but, for long distances, the systems can be very expensive. Like Microwave systems, the attenuation depends heavily on atmospheric conditions and any minor obstructions. The systems must have a high output to overcome intense sunlight.

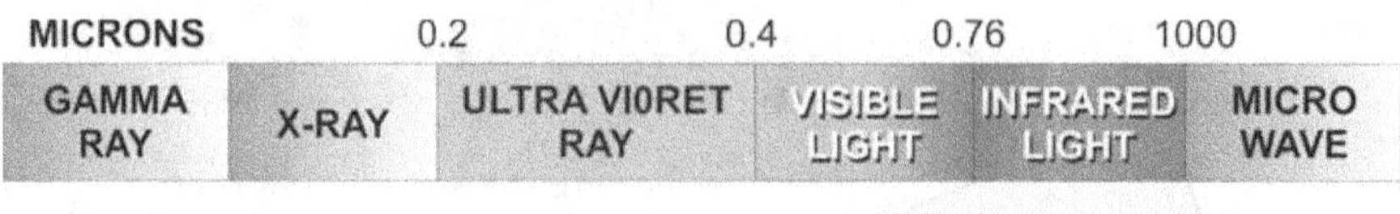

Figure 1-21: Infrared Radiation

Infrared technology may be subdivided into a number of different types as follows:

- **Unidirectional (Point-to-Point)** – These are line-of-sight systems using a focused laser to transmit over distances of a half a mile (kilometer) or more. The careful alignment required by these systems means that the installation could be difficult.

- **Omnidirectional (Broadcast)** – These systems use a 'scattered IR' technology. The light is allowed to reflect off all available surfaces providing an environment for mobile devices. The scattering results in lower data rates of less than 1Mbps and the technology are affected by strong sunlight. These systems are simple to install, as no alignment is required.

- **Reflective** – This system uses optical transceivers located near to devices that transmit the data to a common location for redirection to a receiving device.

Since IR is a light beam, it has some limitations. First, outside light sources such as the sun, certain types of light fixtures, and other devices can interfere with the IR signal. This can also affect the range-most IR remotes have an effective range of 20-30 feet, depending on the environment.

Second, the path between the device and the remote needs to be unblocked. If your gear is not in line of sight with the remote, it will not respond to the signal. This issue is usually avoided by using some type of RF based control system.

Unidirectional (Point-to-Point)

Omnidirectional (Broadcast)

Reflective

TEST TIP

Understand the limitations of IR operations.

Types of IR/RF Commands

When controlling AV gear, there are the two main types of commands:

- **Toggle commands** — Some devices will perform multiple actions from the same command. The most common use of this is the Power button on most remotes. Pressing the Power button turns the TV on or off based on the status of the TV when the command was sent. This concept is illustrated in Figure 1-22. Another toggle command is the TV/Video button, often used to cycle between multiple inputs on a display.

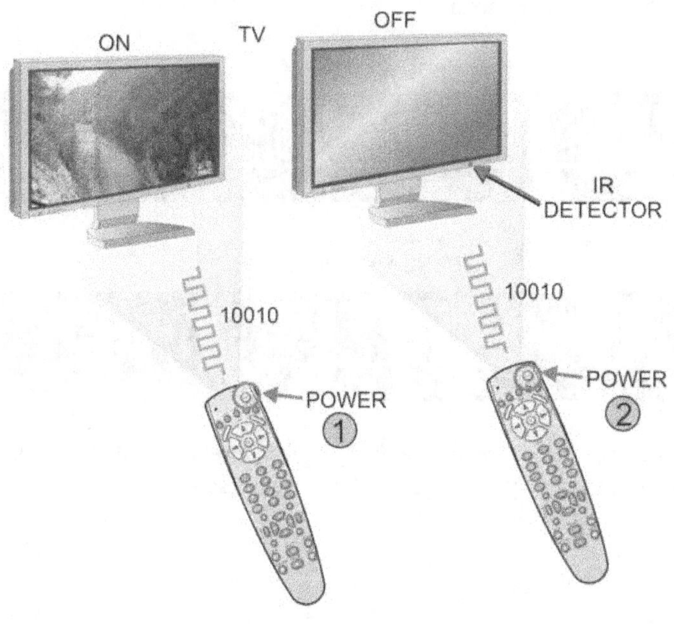

Figure 1-22: Toggle Command

- **Discrete commands** — When using toggle codes, you may have to send a command multiple times to change your settings properly. The use of discrete codes allows specific commands to be sent instead of cycling through multiple options.

  A discrete Power command would mean you have a specific code for On and a different specific code for Off, as depicted in Figure 1-23. No matter what the status of the device, the code will always do the same thing. Sending a discrete On command to a device that is turned on will do nothing.

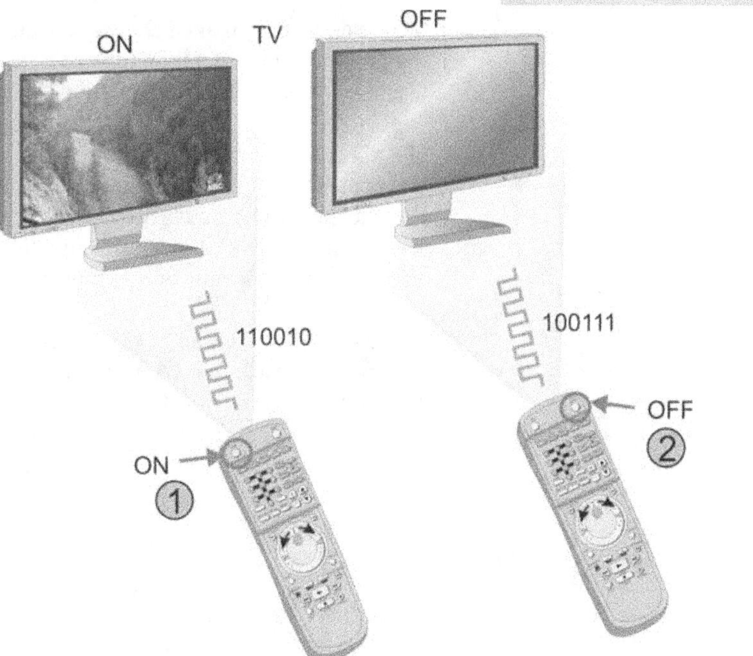

Figure 1-23: Discrete Command

Remote Codes

It is important to remember that **IR codes** are not universal—a Play command for one brand will often not work for other brands. Each manufacturer has their own **codeset**, or group of codes for their own devices.

Figure 1-24 shows a graphical representation of a Brand X power command and a Brand Y power command. You can see the pulses that make up the command, and can see how long it takes to send each code. The pulses are spaced differently, as are the groups. Even though there are two commands with the same function, they are very different IR pulses since they are made by different manufacturers.

Figure 1-24: Power Commands

Many manufacturers do not create an entirely new codeset for each of their devices. Instead, many receivers in a single product line will all use the same codeset.

IrDA

IrDA stands for **Infrared Data Association**. It is an international organization that creates and promotes interoperable infrared data interconnection standards for wireless networks. The name IrDA is associated with products that conform to the interconnection standard.

The IrDA infrared transmission specification allows multiple IrDA devices to be attached to a computer so that it can have multiple simultaneous links to multiple IrDA devices. In these scenarios, the IrDA link provides the high-speed transmission media between the Ethernet devices. IrDA physical specifications require the transmitted signal to be viewed by the receiver up to 1 meter away with the nominal range being 5 cm – 60 cm.

Figure 1-25 illustrates an IrDA-connected printer. The same technology has been employed to carry out transfers between computer communications devices such as modems and Local Area Network cards.

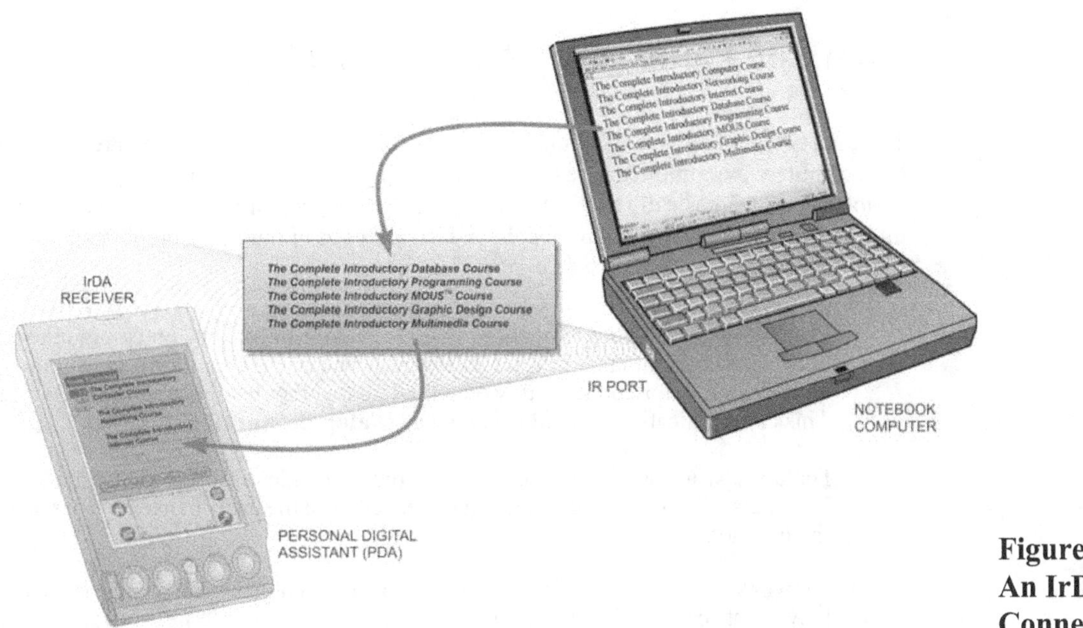

**Figure 1-25:
An IrDA Device
Connection**

Table 1-1 summarizes the specifications for the most widely used serial and parallel interfaces in the residential environment.

PORT/CABLE SPECIFICATION	MAXIMUM LENGTH	DATA TRANSFER RATE
RS-232	50 feet	300 – 9600 bps
USB 1.1 USB 2.0	5 meters 5 meters	12 Mbps 1.5 Mbps, 12 Mbps, and 480 Mbps
Parallel Ports	2 meters	10 Kbps
Enhanced Parallel Port (IEEE 1284)	10 meters	2 Mbps
IEEE1394 (FireWire) IEEE 1394b	4.5 meters 100 meters	100 Mbps, 200 Mbps, and 400 Mbps 800 Mbps
RS-422	4000 feet	10 Mbps
RS-485	4000 feet	10 Mbps
Infrared (IrDA) Ports	2 meters	200 Mbps

**Table 1-1: Standard
Port Cabling Lengths
and Transmission
Speeds**

Common Communication Protocols

There are two general types of communications protocols based on how they are intended to let computers communicate, as depicted in Figure 1-26. These are point-to-point protocols—which basically enable two computers to connect directly to each other to share information—and networking protocols—which are used in multiple computer connection schemes.

- **Point-to-Point Protocols** – These rules are designed to carry on communications between two computers in a direct fashion. The original point-to-point protocols were designed for connecting two computers together using the telephone system. This communication method is known as **Dialup networking**.

  Dialup networking is still widely used to connect residential users to Internet Service Providers (ISPs), which then grant the user access to the many networks that make up the Internet.

- **Networking Protocols** – These protocols are different from dial up protocols in that they must also be able to designate which one of multiple network users is the intended recipient for information being sent across the network media. Devices connected to the network use the protocol to recognize data moving along the transmission media that is intended for it (and ignore information meant for other devices using the network).

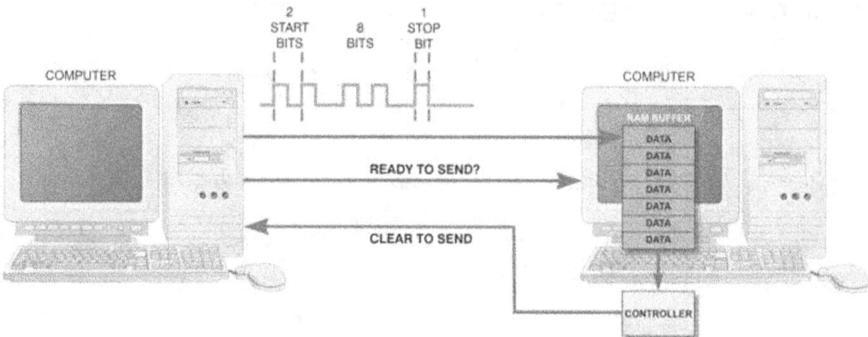

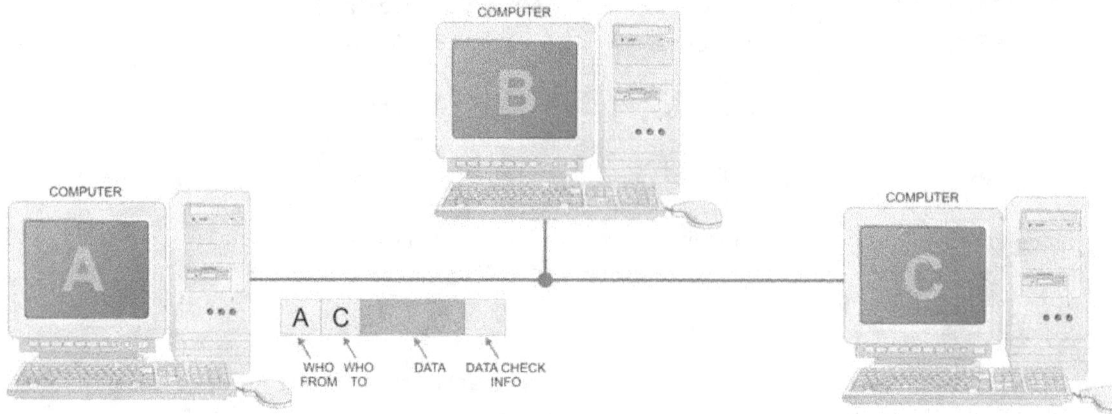

**Figure 1-26:
Communication
Protocols**

Although there have been many different communication protocols developed, one suite of protocols has become the unrivaled leader, the Transmission Control Protocol/Internet Protocol or TCP/IP.

Internet Protocol

All computers, routers, or network-enabled devices that access the public Internet today must be assigned an **Internet Protocol** (**IP**) address. Residential networking allows you to access stored movies, music, and other digital media for use in a dedicated AV system. This is known as **streaming media** from a storage location to the receiver for use. Some receivers also enable the use of Internet Radio—a way to receive radio broadcasts using the Internet.

Some newer receivers are designed so that they can be connected into a home network to share music from a computer to the receiver, as shown in Figure 1-27.

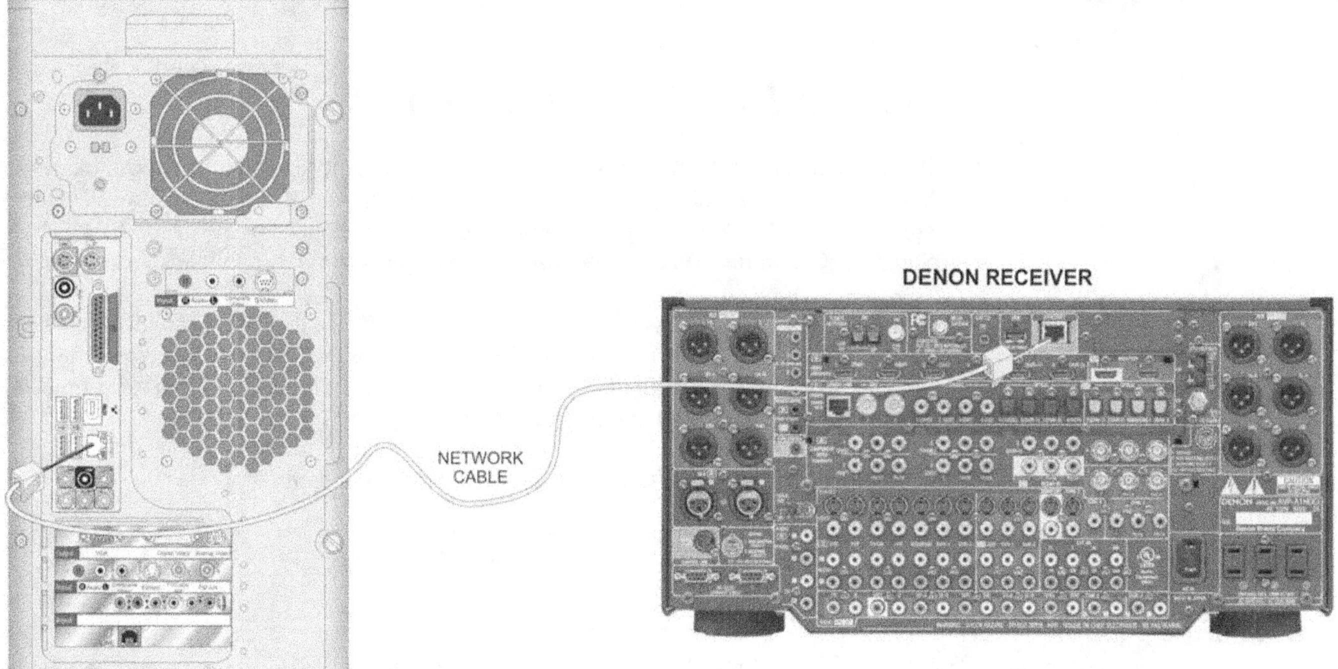

For example, In residential settings local area networks are used to connect different computers and devices together within the home, as shown in Figure 1-28.

Figure 1-27: Adding a Receiver to a Computer Network

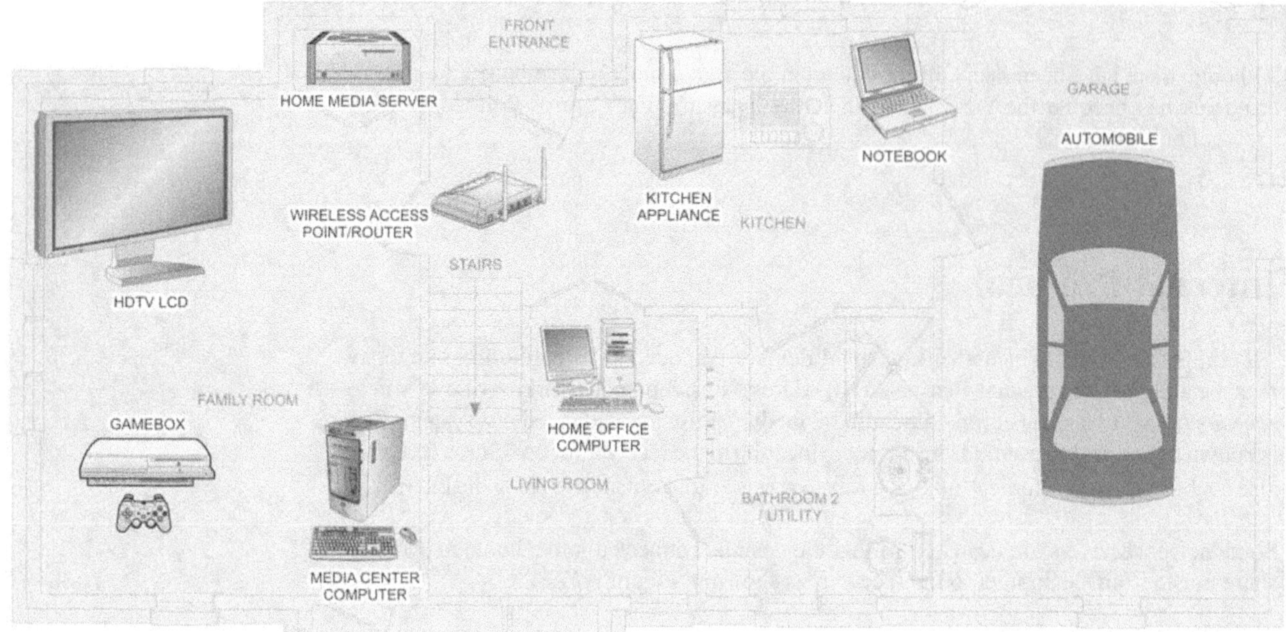

Figure 1-28: Residential Local Area Networking

IEEE 802.11x Wireless Ethernet

Radio frequency (RF) systems use electromagnetic energy to communicate without wires between various control devices. IEEE 802.11 is a popular set of standards for wireless local area network (WLAN) computer communication, developed by the IEEE LAN/MAN Standards Committee (IEEE 802)for use in the 5 GHz and 2.4 GHz public spectrum bands.

WiFi (WIreless-FIdelity) is a logo from the Wi-Fi Alliance that certifies network devices comply with the IEEE 802.11 wireless Ethernet standards. In the early 2000s, Wi-Fi/802.11 became widely used (initially 802.11b, then later, 802.11g/h/I and within a short time, all laptops and other handheld devices came with Wi-Fi built in.

Table 1-2 compares the different IEEE 802.11-family standards.

Table 1-2: IEEE 802.11-Family Standards

NAME	MAXIMUM DATA RATE	FREQUENCY BAND	NUMBER OF CHANNELS
802.11b	11 Mbps	2.4 GHz	11*
802.11a	54 Mbps	5 GHz	12
802.11g	54 Mbps	2.4 GHz	11*
802.11n	Up to 300 Mbps	2.4 GHz (5 GHz also supported on some devices)	11*, 12#

* Only channels 1, 6, and 11 do not overlap with each other
\# Also supports channel bonding for greater throughput if it will not interfere with other networks

Bluetooth

Bluetooth wireless technology is designed to connect one device to another with a short-range radio link. This standard evolved from early work and engineering studies performed by Ericsson Mobile Communications in 1994. The Bluetooth project focused on the feasibility of designing a low-power/low-cost radio interface between mobile phones and their accessories.

A Bluetooth radio consists of an RF transceiver, a baseband link control with associated link management software, and an antenna subsystem. The radio uses frequency-hopping, spread-spectrum technology to support both point-to-point and point-to-multipoint connections.

Under the current specification, up to eight Bluetooth-enabled devices can automatically configure themselves into a "piconet," with one master and seven slaves. Although a piconet can comprise no more than eight devices, its coverage can be extended by attaching one of the slaves to other piconets. Most Bluetooth implementations support a range of up to 10 meters and speeds up to 700 kbps for data or isochronous voice transmission.

Radio frequency operation is in the unlicensed industrial, scientific, and medical (ISM) band at 2.4 to 2.48 GHz, using a spread-spectrum, frequency-hopping, full-duplex signal at up to 1600 hops/sec.

The signal hops among 79 frequencies at 1 MHz intervals to give a high degree of interference immunity. The range of each radio is approximately 10 meters, but can be extended to around 100 meters with an optional amplifier. RF output is specified as 0 dBm (1 mW) in the 10 m range version and -30 to $+20$ dBm (100 mW) in the longer range version.

Bluetooth wireless technology eliminates the need for numerous, often proprietary, cable attachments for connection of practically any kind of communication device. Connections are instant and are maintained even when devices are not within line of sight.

The key parameters associated with the Bluetooth specification are shown in Table 1-3.

CHARACTERISTIC	SPECIFICATION
Frequency Band	2400 – 2483.5 MHz (ISM band)
Modulation	Gaussian-filtered binary frequency-shift keying at a line rate of 1 Mbps
Frequency Hopping Rate	1600 hops/sec in normal operation with four special hopping sequences reserved for connection setup
Transmit Power	Class 1; 1–100 mW (power control required) Class 2; 0.25–2.5 mW Class 3; 1 mW
Range	0.1 to 10 meters or up to 100 meters power Class 1

Table 1-3: Bluetooth Specifications

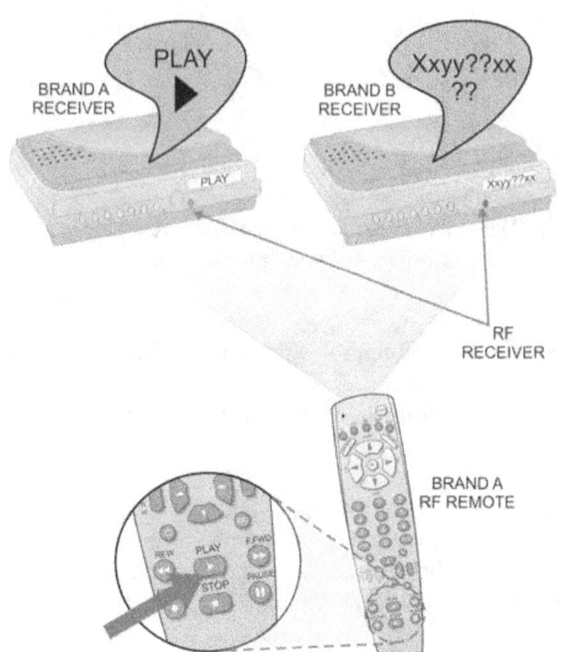

**Figure 1-29:
IR Receivers and
Remote Frequencies**

RF controllers employ radio waves instead of light pulses to transmit commands. Because RF can pass through walls and is less susceptible to interference than IR, it is often used in applications when an IR based control system is inadequate. If a device is built to accept RF commands, then the device and the remote must use the same RF frequencies. This usually means they will be built by the same manufacturer, as shown in Figure 1-29.

A more common approach is to use an RF/IR repeater. This is a base station that takes RF commands from a remote, and emits IR codes out to the devices. Because most AV equipment is IR based, this allows a much wider variety of equipment to be used while taking advantage of the RF technology.

Security Protocols

Recall that in security systems, all sensors appear to the system as a switch. The individual input devices monitor their specified variable conditions and convert that information into a switch-open or switch closed input signal. This standard (protocol) is referred to as **Contact Closure** and is used to monitor conditions such as open door, temperature limit reached, power failure occurrences, or intrusion detection.

Most security controllers have a finite number of input channels that can be used to cover all the security monitoring points for the residence. This usually requires that the residence's various input devices be grouped into logically related input zones.

While security zones are created by physically wiring devices together and attaching them to one of the controller's input terminals, the zone parameters can be defined during the zone programming process. A zone can be programmed to function as a normally closed (NC) input where all the monitoring devices provide a closed circuit when they are not activated, or as a normally-open (NO) input where the devices function as an open circuit until they are activated.

If a zone is configured to function as a normally open zone, the input terminal does not sense any current flow through the input until the sensor is activated. Devices in NO zones are wired in parallel. On the other hand, a zone configured to operate as a normally closed zone has a constant current flow through the terminal until one of the sensors is activated (opened). Devices in a NC zone are wired in series with each other. It is important that the controller's input type be configured so that it is compatible with the type of zone being created.

Most security systems enable the installer to create descriptive visual output messages that correspond to the nature or location of the alarm condition. For example, if the installer has configured a zone consisting of all the west side windows, he may create a corresponding visual message that says "Trouble At West Side Windows". This message is generally constructed using ASCII coded characters entered in the controller's "Zone Message" field.

ASCII

ASCII is an acronym for American Standard Code for Information Interchange. The ASCII code was originally developed for teletypewriters but eventually found wide application in personal computers. The standard ASCII code uses seven-digit binary numbers. The code can represent 128 different characters, since there are 128 different possible combinations of seven 0's and 1's. The binary sequence 1010000, for example, represents an uppercase "P," while the sequence 1110000 represents a lowercase "p."

Many security systems also employ automatic dialers with voice capabilities to communicate with remote monitoring stations. In some systems, the dialer allows the installer to pre-record actual voice messages that the system can play back for an associated condition when the phone is answered. Other voice systems enable the installer to create phonetic messages using ASCII coded messages that play through the voice generator circuitry.

Many security systems also employ automatic dialers with voice capabilities to communicate with remote monitoring stations. In some systems, the dialer allows the installer to pre-record actual voice messages that the system can play back for an associated condition when the phone is answered. Other voice systems enable the installer to create phonetic messages using ASCII coded messages that play through the voice generator circuitry.

Lighting Control Protocols (Open Standards)

The home control protocols, including some of those also covered earlier are also used for home lighting control. The lighting control protocols are:

- Z-wave
- Zigbee
- Powerline carrier (X-10 protocol /PLC)
- UPB (Universal Powerline Bus)

Z-wave and Zigbee

Z-wave and **Zigbee** are two short-range wireless communication protocols similar to Bluetooth. Both are designed to control home automation devices using RF technology.

Z-wave was developed by the Danish company Zensys and the Z-Wave Alliance. It is a RF-based, two-way **mesh networking** protocol that allows for the control of thermostats, garage door openers, light switches, and residential appliances.

One of the great advantages of Z-wave's technology is its mesh networking capability that allows for the wireless signals to be routed from device to device until the target device is reached, as illustrated in Figure 1-30. This enables users to control their first floor lighting from the driveway by sending the appropriate command from their remote to the garage door opener, which in turn forwards the command to the next closest node. Users will also receive a confirmation for the issued command, which will be routed back to the user's remote control in the same manner.

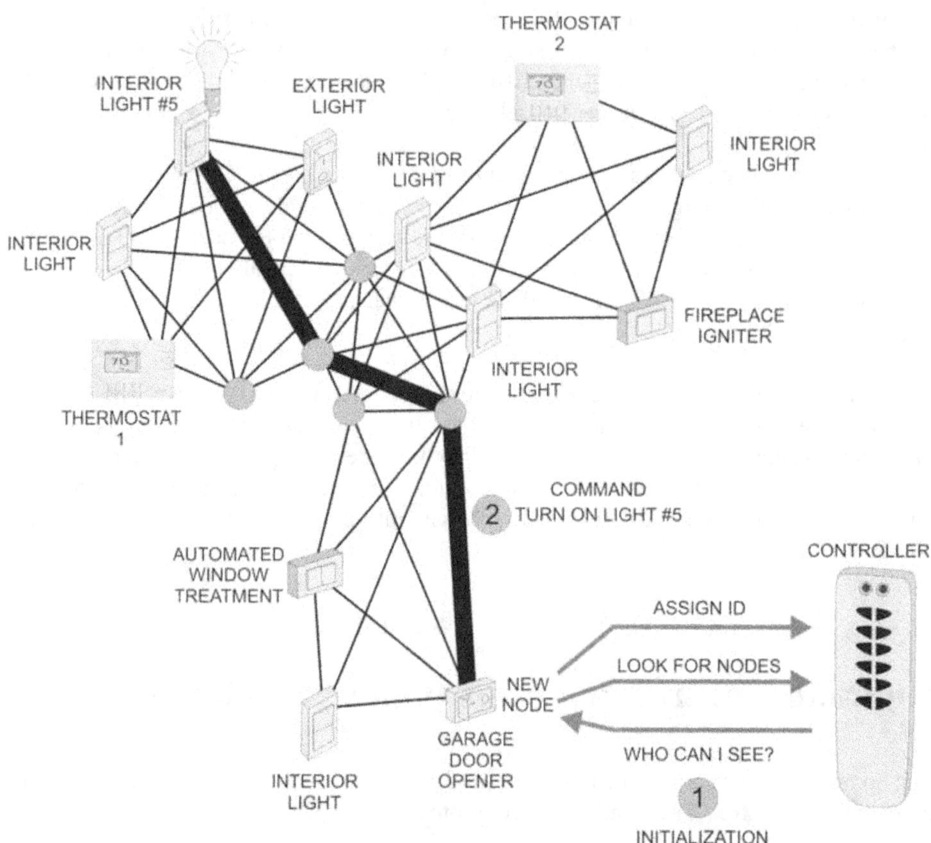

**Figure 1-30:
Z-wave**

The Z-wave radio uses the 900 MHz ISM band: 908.42MHz (USA); 868.42MHz (Europe); 919.82MHz (Hong Kong); 921.42MHz (Australia/New Zealand). Z-wave supports the control of up to 192 devices. With Z-wave, it is possible to control indoor and outdoor lighting. The Z-wave protocol has a range of up to 100 meters and routes signals from one device to another making it possible to have the network as large as you need.

Zigbee is a similar specification for a wireless communication protocol using small, low-power digital radios based on the IEEE 802.15.4 standard for **wireless personal area networks (WPANs)**. An example of a Zigbee application is to connect wireless headphones with cell phones via short-range radio. Zigbee also supports the control of lighting, home automation, security systems and industrial control.

It is a very fast protocol with the ability to control up to 64,000 nodes from a single controller. The wireless Zigbee network can be configured in a variety of different topologies, as illustrated in Figure 1-31. The technology is intended to be simpler and cheaper than other WPANs, such as Bluetooth. Zigbee is targeted at radio-frequency (RF) applications that require a low data rate, long battery life, and secure networking.

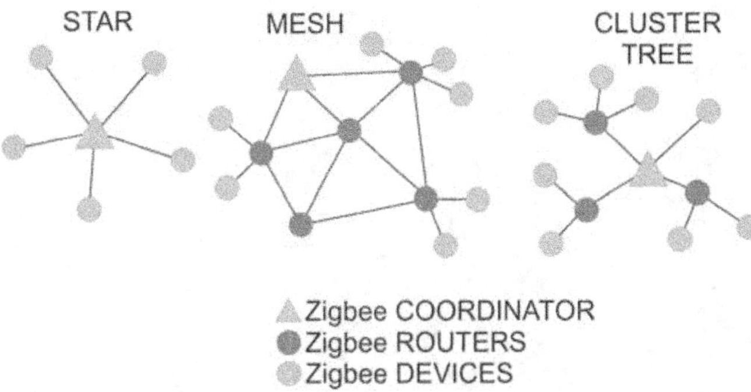

Figure 1-31: Zigbee Network Topologies

Zigbee also operates in the industrial, scientific and medical (ISM) radio bands; 868 MHz in Europe, 915 MHz in countries such as USA and Australia, and 2.4 GHz in most jurisdictions worldwide. Zigbee chip vendors typically sell integrated radios and microcontrollers with between 60K and 128K flash memory. Zigbee uses spread-spectrum signaling in three different frequency ranges, as shown in Table 1-4.

Table 1-4: Zigbee Specifications

DATA RATE	FREQUENCY RANGE	MODULATION	CHANNELS	CHANNEL NUMBERS	LOCATION WHERE USED
250 Kbps	2.4 GHz ISM	O-QPSK	16	11-26	Worldwide
20 Kbps	868-870 MHz		1	0	Europe
40 Kbps	902-928 MHz (aka 915 MHz)	BPSK	10	1-10	Americas, Pacific Rim

Power Line Carrier

Power line carrier (PLC) is a communication technique that uses the existing power wiring to carry information, as depicted in Figure 1-32. All power line carrier systems operate by impressing a modulated carrier signal on the wiring system. Different types of powerline communications use different frequency bands, depending on the signal transmission characteristics of the power wiring used. Since the power wiring system was originally intended for transmission of AC power, the power wire circuits have only a limited ability to carry higher frequencies. The propagation problem is a limiting factor for each type of power line communications.

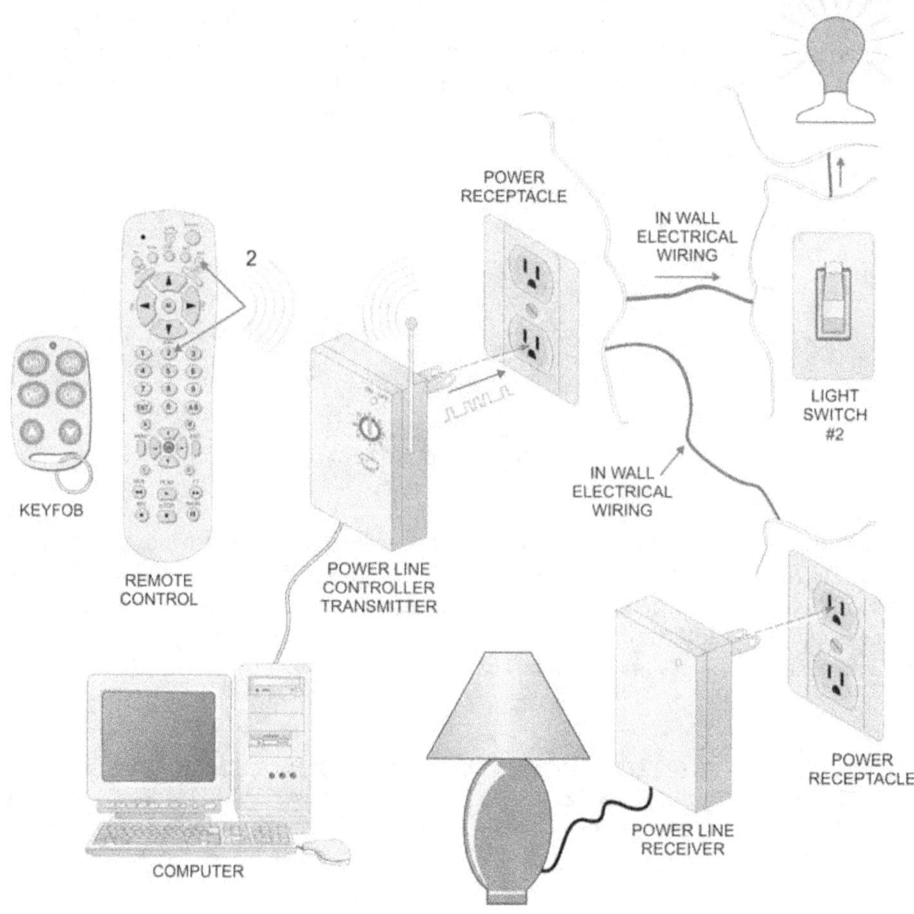

Figure 1-32:
Power Line Carrier
Communication
System

X-10

X-10 is a popular standard PLC signaling and home control protocol language that enables signals to be transmitted anywhere in the home over existing 120-volt electrical lines. Controllers are installed by simply plugging in a controller (transmitter) at one electrical outlet and plugging in a module (receiver) at any other outlet.

In the 120Vac 60Hz power scheme, a 1-millisecond burst of 120 kHz at the zero crossing point represents a binary digit 1. In a perfect world, the zero crossing point would occur at 0°, but allowances for real world imperfections permit timing variations within 200 microseconds of the zero crossing point. The immediate absence of a pulse following this 120kHz burst serves to legitimately identify the previous pulse. Alternately, the absence of a 120kHz pulse at the zero crossing points, immediately followed by the presence of the pulse, signifies a binary digit 0.

X-10 transmits data packets at an effective data rate (fewer overhead bits) of 18.62 bps during the zero crossing time of the 60 Hz alternating current waveform on the power line. The address for each device can be manually selected during the installation of the modules. This permits remote control of any device plugged into the receiver from any location.

Each signal (message group) is transmitted two times for purposes of redundancy, reliability, and to accommodate line repeaters. This practice also helps to reduce the chances of false signaling, and ensures that receivers will recognize the transmit data above the analog noise normally being generated on the power lines. Taking into consideration the additional time required for retransmission and line control, X-10 data transmission is too slow for any technology more complicated than turning various devices ON and OFF.

A predictable starting point for a data frame is accomplished by first sending a **start code** of 1110 (pulse, pulse, pulse, absence of a pulse). Following this, a **letter code** (also called a house code, from A to P) is transmitted. Then, the **function code**, consisting of either a unit number code (1 to 16) identifying which X-10 unit should respond is transmitted, or a command code, identifying the specific action to perform. The last bit in the function code determines whether it is unit identifier (0) or a command (1). Therefore, an X-10 **frame** consists of one start code, one letter code, and one function code. A complete power line cycle consists of two complete frames, permitting the complete data block, (start code, house code, function or key code to be transmitted in groups of two. Bright and dim lighting commands are exceptions to this rule, and should be transmitted continuously (at least twice) with no gaps between codes.

In order to identify that data is either targeting a new address, switching a function from an address to a command, or sending more than one command, the subsequent frame must be separated by a sequence of at least 6 clear zero crossings (000000). This translates to three power line cycles between each group of two codes and serves to reset the shift registers in all devices. Within each block of data, each four or five bit code should be transmitted in true compliment form on alternate half cycles of the power line. This means that if a 1-millisecond burst of signal is transmitted on one half cycle (binary 1) then no signal should be transmitted on the next cycle, (binary 0). Also, keep in mind that 'negative logic' is being used to implement these codes.

The relatively high-frequency carrier frequency carrying the X-10 signal cannot pass through a power transformer or across the phases of a multiphase system. For split phase systems, the signal can be passively coupled from phase-to-phase using a passive capacitor, but for three phase systems or where the capacitor provides insufficient coupling, an active X-10 repeater can be used. To allow signals be coupled across phases and still match each phase's zero crossing point, each bit is transmitted three times in each half cycle, offset by 1/6th cycle.

Universal Powerline Bus

Universal powerline bus (UPB) is a home control system that uses the existing power wiring to transmit control information between UPB devices. UPB pulses are generated by charging a capacitor to a high voltage and then discharging that capacitor's voltage into the powerline at a precise time. This quick discharging of the capacitor creates a large "spike" (or pulse) on the powerline that is easily detectable by receiving UPB devices wired large distances away on the same powerline.

While transmitting, one UPB pulse is generated each half-cycle of the 60Hz AC electrical power cycle. The generation of each UPB pulse is precisely timed to occur in one of four predefined positions in the half-cycle of the AC powerline. The position of each UPB pulse determines its value as either 0, 1, 2, or 3. This method of encoding data as a relative position of a pulse is a well-known and used method in digital communications known as **Pulse Position Modulation (PPM)**. Since each UPB Pulse can encode two bits of digital information and there are 120 AC half-cycles per second (at 60Hz), UPB communication has a raw speed of 240 bits per second. Although this speed isn't fast enough for doing high bandwidth applications it is perfectly adequate for doing command and control communication.

UPB pulses are transmitted in a special region toward the end of the AC half-cycle known as the **UPB Frame**. This region was selected due to its relatively low noise characteristics and for other attributes that make it an optimum position for powerline communications. UPB Frames are synchronized to the low-to-high transition of the AC waveform (known as the AC zero-crossing point) such that one Frame starts T/Frame microseconds after the zero crossing and the other Frame starts 8,333 microseconds (one half-cycle at 60Hz) after the first one.

UPB controllers range from extremely simple plug-in modules to very sophisticated whole house home automation controllers. The simplest controllers are plug-in controllers that are recommended for a moderate amount of switches and devices as it becomes cumbersome to control a wide range of devices. More sophisticated controllers can control more units and/or incorporate timers that perform pre-programmed functions at specific times each day. Units are also available that use passive infrared motion detectors or photocells to turn lights on and off based on external conditions.

Finally, whole house home automation controllers can be fully programmed. These systems can execute many different timed events, respond to external sensors, and execute, with the press of a single button, an entire scene, turning lights on, or establishing brightness levels.

Proprietary Protocols

Proprietary protocols are often referred to as "**closed**" **protocols** that are not "open" or shared freely among all users. As an example, the IEEE family of approved standards are considered **open protocols**. Proprietary RF and low voltage protocols usually are patented or licensed for use by a sponsoring corporation and are provided to user organizations by licensing agreements.

There are several proprietary systems for managing and controlling home automation systems. These systems may provide a wide variety of capabilities that are not found in standard products, however, remember that these devices will only work with the single manufacturers products

Proprietary systems are at a disadvantage in that they may not be able to interoperate with standardized products. There is also some risk associated with the potential for interference that may occur in the same environment between proprietary systems and standard products. With standard products like X-10 and UPB, this risk is not usually an issue.

HVAC (HEATING VENTILATION AND AIR CONDITIONING) TERMINOLOGY

As a home technology integrator you will be expected to know how to identify the various HVAC design concepts, components, and control systems. You should also become familiar with programming thermostats and performing basic maintenance and troubleshooting tasks.

Installation, maintenance and troubleshooting are normally performed by licensed HVAC technicians. However, you should be able to describe the basic maintenance and troubleshooting tasks that can be accomplished by a homeowner or RESI technician.

The following section describes the HVAC control and communications layers and also identifies the terms you will need to know for the RESI Environmental Control Certification exam.

HVAC Concepts

HVAC designs for automated homes now include more than the standard heating and cooling units familiar to most home designers. An automated home may include zones for control of heating and air conditioning settings in different areas of the home.

The temperature in each zone is controlled by using multiple thermostats and motorized dampers installed in the ducts.

In large homes, multiple distributed heating and cooling units may be required to match the floor plan and HVAC distribution ducts. More traditional designs dictate a single centralized heating and cooling unit.

HVAC systems are installed by trained professionals who are familiar with building codes as well as fire and safety regulations. As a home integrator, you will need to become familiar with the terms; the equipment categories; and the functional relationships between such items as programmable thermostats, zone controls, ducts, whole-home fans, damper controls, condensers, evaporators, furnaces, air handlers, and heat pumps.

Zoned/Non-Zoned Designs

Zoned heating and cooling allows customized settings for temperature in specific areas of a home. **Non-zoned systems** regulate the complete home from a single control point.

Zoned system designs should be considered when the residence is multilevel, or is designed with wings or sprawling ranch-type styling. In large open areas with vaulted ceilings or rooms with many windows zoning would also be useful. A zoning system would also be useful in a home with a finished basement or attic, an indoor hot tub or swimming pool, or an exposed concrete floor.

Zoned Design Considerations

When a zoned system is installed, the communicating thermostats located in various areas of a home feed information to the automated zoning panel.

The zoning panel sends commands to open and close dampers located in the heating and cooling ducts to control the temperature in each room and thereby eliminate hot and cold spots in various areas of a residence.

When used with programmable thermostats, a zoning system can save up to 30% on heating and cooling costs. Each zone can be customized for its own climate, controlling the temperature differences between floors, keeping the bedrooms cooler at night, or adjusting the living room temperature to account for large windows.

After the zones have been established, each zone must be provided with its own sensor or thermostat.

Each zone is also given its own control mechanism. In a forced air system this mechanism is typically a motorized damper that is used to control the volume of warm or cold air that flows into the zone.

When a thermostat in a particular zone detects a need for heating or cooling, it transmits a signal to a centralized control panel. The control panel manages the activities of all the zones by driving the motors that control the dampers in the duct work as shown in Figure 1-33. It causes the damper motor to open or close the zone damper depending on the demands of the zone thermostat.

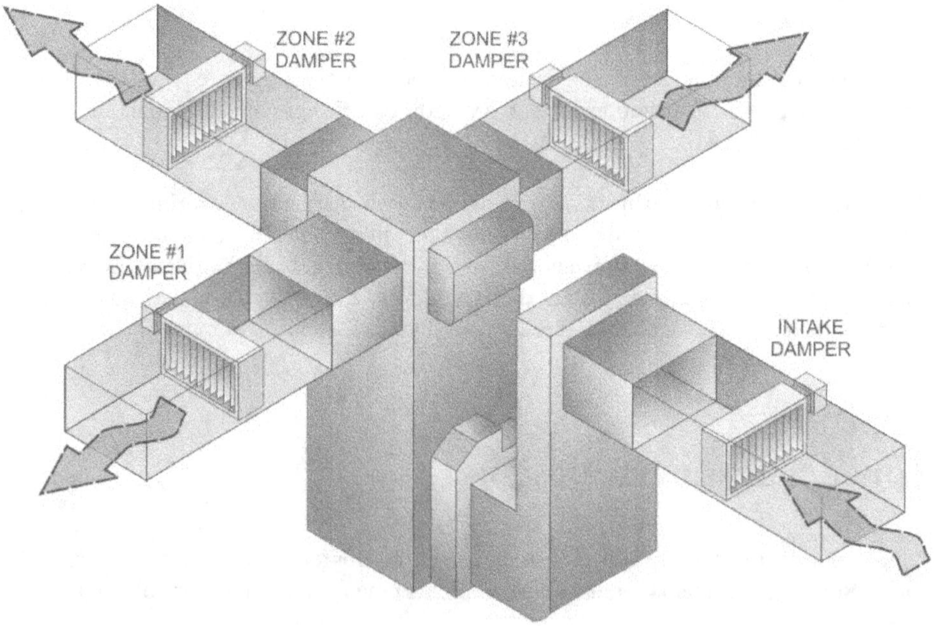

Figure 1-33: Duct Work with Motorized Dampers

If a particular zone requires more heat, the controller will drive the motor to open the damper to that zone, allowing more heated air to flow into the zone. It also closes the dampers for all the other zones. In this manner, the full heating/cooling energy of the system is directed into the one zone until the specified thermostat indicates that the desired temperature has been reached.

TEST TIP

Know the damper requirements for individual HVAC zones.

TEST TIP

Remember that all zones and corresponding dampers are run by a single controller.

TEST TIP

Understand the relationships between multiple thermostats and a damper controller.

HVAC CONTROL SYSTEMS

This section focuses on the types of control and management systems associated with residential HVAC systems. The topics include:

- Thermostats

- Staging thermostats

- Electronic thermostats

- Damper controls

- HVAC wiring and termination points

- Zone control distribution panels

Thermostats

Thermostats are devices that measure the temperature of the surrounding air. When the temperature deviates from a user-selected temperature setting, the thermostat sends a signal to the furnace or air conditioner to turn on.

Thermostat designs range from the older "legacy" type thermostats that typically do not have electronic circuitry, to intelligent thermostats with numerous programmable features.

Older manual control thermostats occasionally use **bimetal** contacts that react to temperature changes by changing their physical shape, creating an electrical switch contact closure, or opening, depending on the temperature variation. They use a rotary temperature selector similar to the type shown in Figure 1-34.

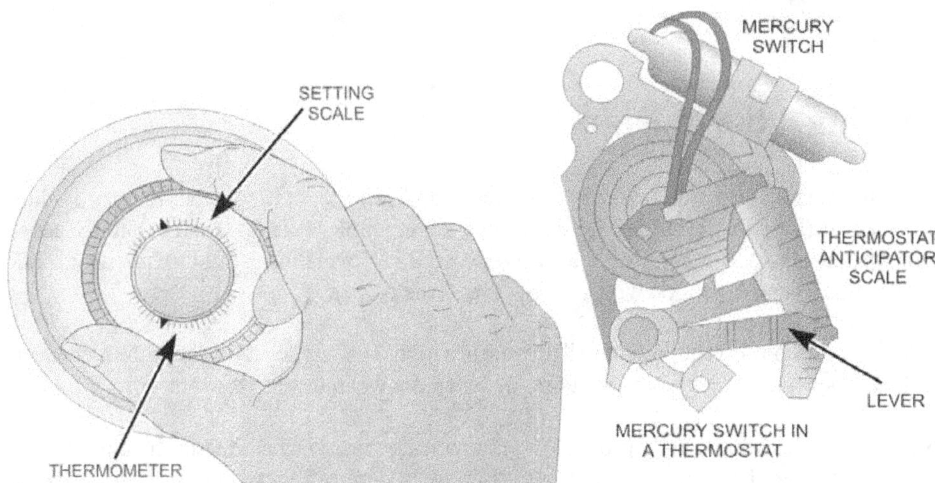

Figure 1-34: Manual Thermostat

Legacy manual thermostats that incorporate a mercury switch must be mounted completely horizontal so that the liquid mercury can use gravity to make and break the contact between its embedded wires. Mercury thermostats are being phased out and are no longer used due to the toxic nature of mercury. Programmable electronic thermostats are mercury-free and are more energy-efficient than the mercury type manual thermostats.

Line Voltage Thermostats

A **line voltage thermostat** is normally used with baseboard or radiant heating systems. It is uncommon for a line voltage thermostat to be used for cooling chores, although these device types are available. However, a line voltage unit is only to be used for either heating or cooling, rather than for both. Although legacy line voltage models used 120 volts, modern line voltage thermostats use direct connections to 240 volts with larger wires, and are designed to carry the full current drawn by the heater. Line voltage thermostats are not to be confused with low voltage thermostats, or used with gas heating systems. Keep in mind that trying to install or connect the wrong thermostat can cause fires and other accidents. Always be sure the matching system has been installed for the selected line voltage thermostat.

TEST TIP

Know when an HVAC system requires the use of a line voltage thermostat.

Electronic Thermostats

Modern HVAC systems are generally equipped with smart electronic control systems. Smart thermostats connect into the HVAC system via the existing residential wiring and communicate with the various control modules through three-conductor control cables. They often include innovative features such as large and easy-to-read displays, EnergyStar-compliant circuitry, manual override for all functions, and hold buttons to prevent automatic operation. They also permit the alternate display of the outdoor temperature and the indoor temperature.

Programmable thermostats have all the characteristics of digital thermostats, but they allow the user to set up timed setting changes. Programmable thermostats have a built-in clock. You can program them when to change settings. Most programmable thermostats allow up to 4 temperature changes per day, and can have a different program for each day of the week.

Staging Thermostats

Staging thermostats provide greater efficiency and lower operating expense by providing multiple levels of heating and cooling capacity. For example, a staging thermostat will cause a multistage HVAC system to operate at a reduced BTU level during mild weather, and shift into a higher capacity mode when the weather turns colder.

Likewise, the cooling equipment will operate at a higher efficiency level when the weather is mild, and shift into a higher capacity mode when the weather becomes hotter.

Staging thermostats typically provide an automatic changeover feature that enables the unit to change over from a heating thermostat to a cooling thermostat without human intervention.

The homeowner can simply set the desired heating and cooling temperatures for the residence, or zone, and the system maintains the residence at those points regardless of the season. To avoid creating a situation where the heating and cooling units are working against each other, the RESI Environmental Control technician should remember that the cooling setting should not be set below the heating set point and the heating setting should not be set above the cooling set point.

Most thermostats include a fan switch that permits the user to circulate air throughout the system without turning on the heating or cooling units.

When this switch is set in the AUTO position, the fan runs in conjunction with the heating and cooling systems. However, when the switch is set to the ON position, the fan runs continuously until returned to the AUTO position.

The so-called "smart thermostat" shown in Figure 1-35 provide additional features mentioned in the previous paragraphs such as staging, time-of-day programming, and seasonal presets.

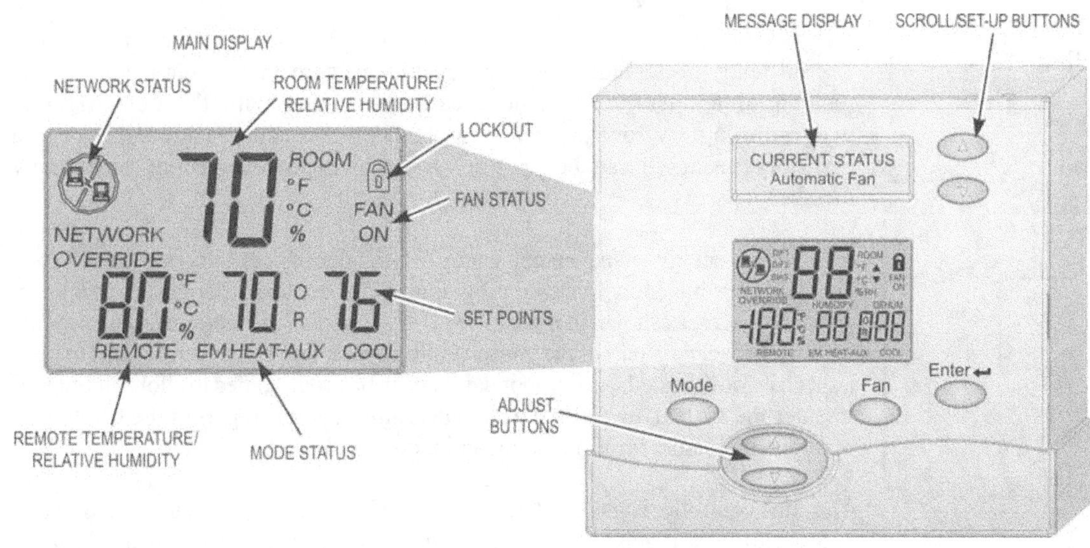

Figure 1-35: Programmable Intelligent Thermostat

Creating HVAC Zones

Zoning a HVAC system can help save energy as well as make a home more comfortable. Multiple HVAC units, each serving a different area or floor are not zoned; they are completely independent systems. **Zoning** is where dampers are used to direct heating and cooling from a single HVAC system to more than one area, as needed. A two-zone system will normally have two dampers, each one controlling the airflow to a zone (which can be multiple registers.) The zone control system will control the dampers and HVAC system to heat and cool each zone.

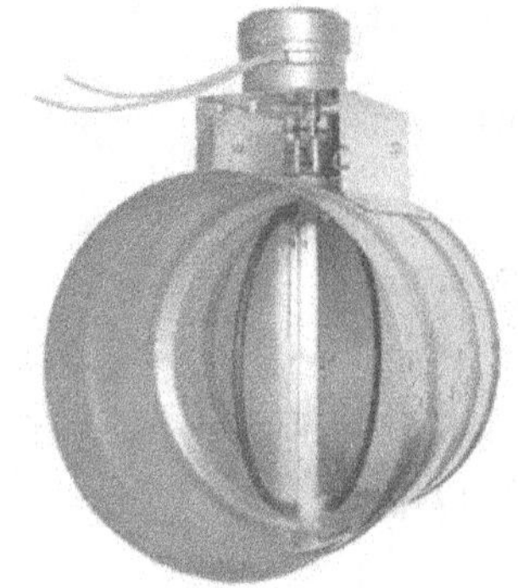

Figure 1-36: Motorized Damper

Damper controls are movable plates located in the distribution ductwork that regulates the airflow. Dampers can be controlled with motors, as mentioned earlier in this book, or manually as discussed here. A motorized damper is shown in Figure 1-36. They are typically used in a zoning application to direct air to the areas that need it most. To accomplish this, they often limit the flow of heated/cooled air to the required areas by shutting off airflow to unoccupied or unused rooms. Due to the increased airflow in the occupied areas, the temperature changes rapidly, resulting in shorter run times by the heating/cooling system.

Certain rooms are habitually warm or cool, regardless of the settings at the thermostat. If some rooms are too hot or cold, the problem may be cured by adjusting the dampers in the outlet registers that supply air to the problem room.

HVAC Zone Programming

The difficulty of the family reaching unanimous agreement about a comfortable residential temperature can be greatly alleviated with the help of **zone programming**. By zone programming the HVAC system, the ground floor of a two-story residence can be properly heated without overheating any family members upstairs.

Residential energy management using a typical high-capacity integrated zone controller permits individualized programming of up to six thermostats, dampers, or water valves, as shown in Figure 1-37. For example, one thermostat can provide cooling for a certain part of the house, while another thermostat can be used to keep a specific room heated. An outdoor damper can be programmed to allow outdoor air to enter the system, or an outdoor temperature sensor can permit the user to see remote temperature information on an indoor thermostat.

The zone controller is mounted near the HVAC equipment or in a utility closet and appears to be a regular thermostat to the existing HVAC equipment. Running properly with as few as two zones, the system requires a wall thermostat and one normally open duct damper for each. The user can add more zones as the residential requirements grow.

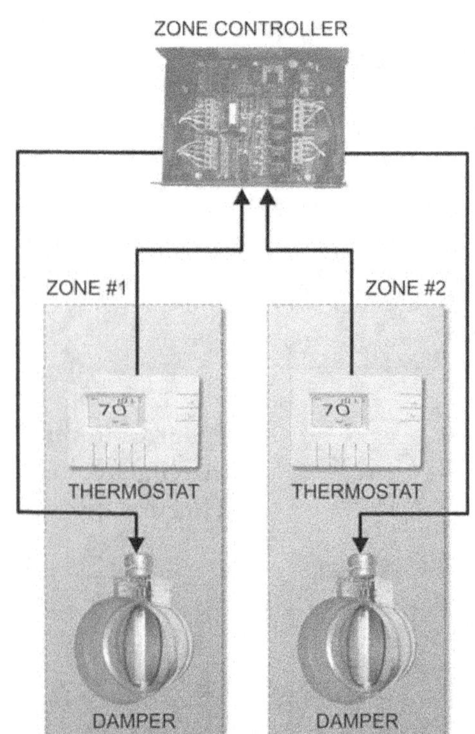

Figure 1-37: Zone Programming

Master/Slave Thermostat Configuration

Thermostats can be programmed to function in a master/slave relationship, as illustrated in Figure 1-38. One thermostat can be installed in the zone designated as the **master zone**. Additional thermostats can be connected using RS-232 serial interface cables. Control information can be programmed into the master thermostat or from a separate microprocessor depending upon the requirements defined in the master thermostat specifications for control of the slave thermostats in a multizone configuration.

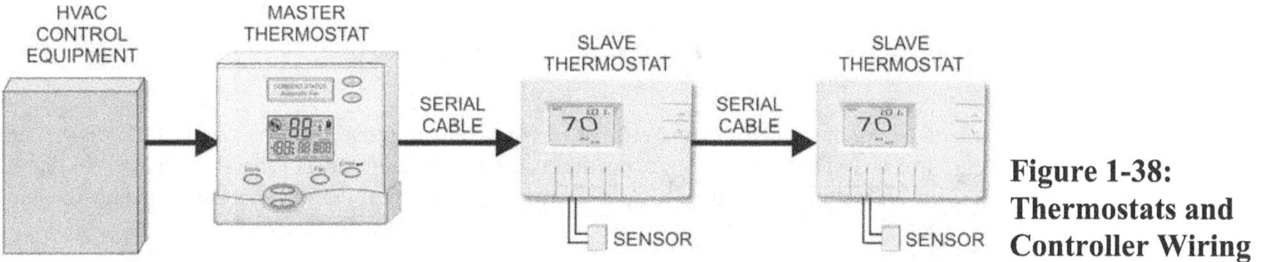

Figure 1-38:
Thermostats and
Controller Wiring

Zone Control Distribution Panels

An HVAC zoning control panel can manage the demands of one zone calling for cooling while another calls for heating. It automatically sorts out these conflicting demands and runs the heating or cooling equipment to provide optimum comfort in all zones.

As shown in Figure 1-39, the zoning control panel is positioned between the HVAC equipment (the furnace/air conditioner and zone dampers) and the various zones' sensing and input devices (temperature sensors and thermostats). The diagram also defines two distinct connection types—those made at the **control layer** and those made at the **communications layer**.

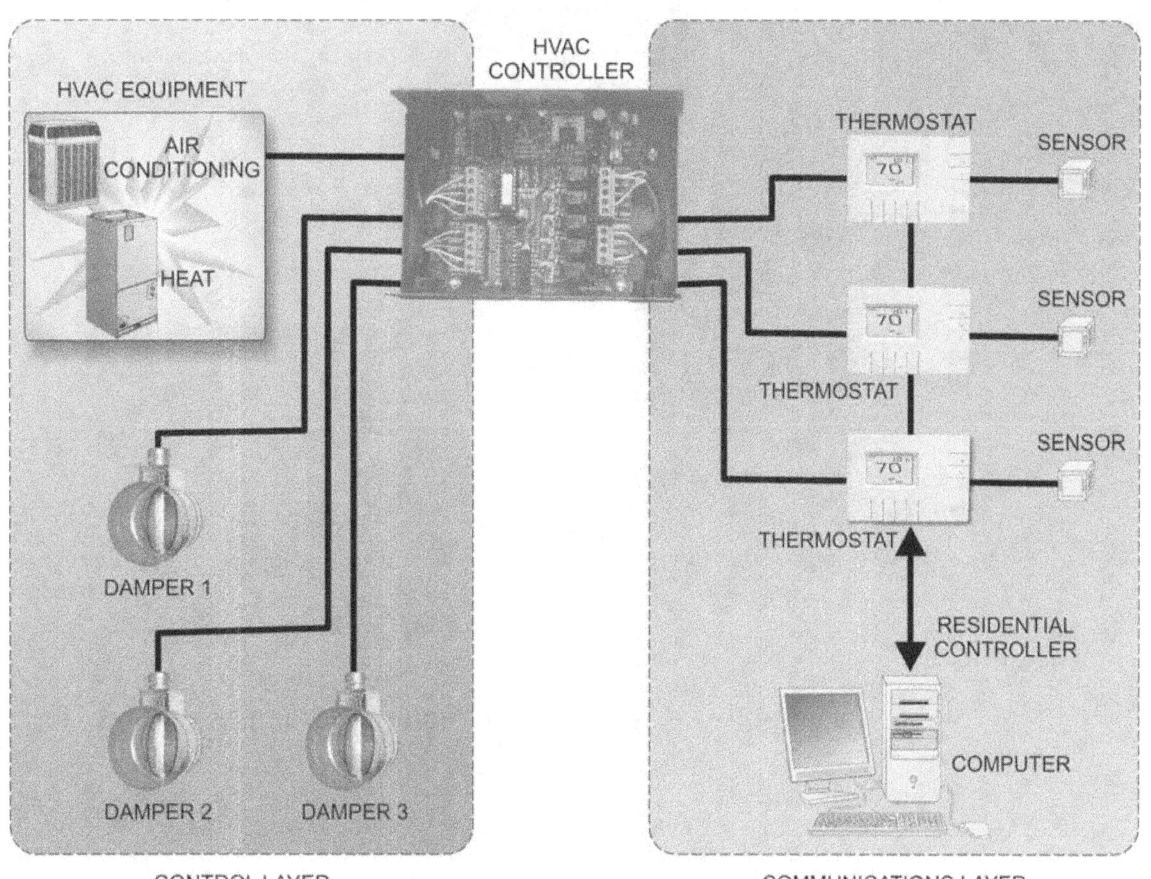

Figure 1-39:
Zone Control
Distribution
Panels

The Control Layer

The connections of the control layer must match those used by the HVAC equipment industry. This typically involves the connection scheme and color code described in Figure 1-40. It is important that the specifications for the controller match those of the heating and cooling components.

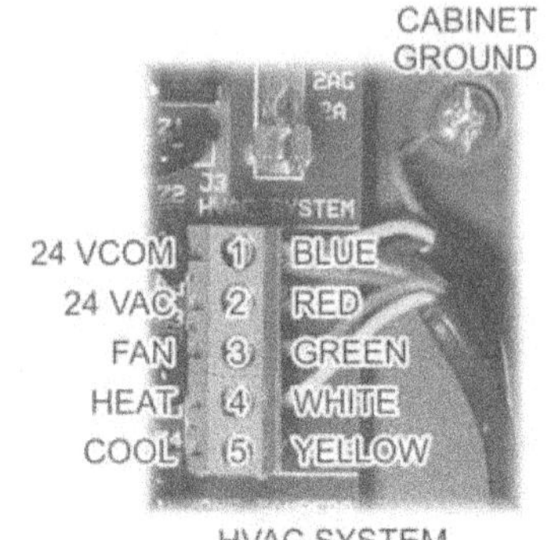

Figure 1-40: HVAC System Connections

Know what voltages and AWG sizes are used with HVAC equipment.

In this example, the controller shares a 24 Vac supply with the HVAC equipment. It also provides three separate control lines to the HVAC equipment using 18 AWG wire—heat, fan and cool. Other equipment may require additional lines to control the operation of compressors and other HVAC related equipment. Since most thermostats are designed so that they can be connected directly to the HVAC equipment in a single zone, stand-alone environment, they also provide similar connections and use the same color codes.

Dampers are also connected to the controller as shown in Figure1-41. These connections are basic two-wire 24Vac signal connections to each damper motor. However, there is no industry standard color-coding for these connections—each damper should have a different color pair to aid in any future troubleshooting operations.

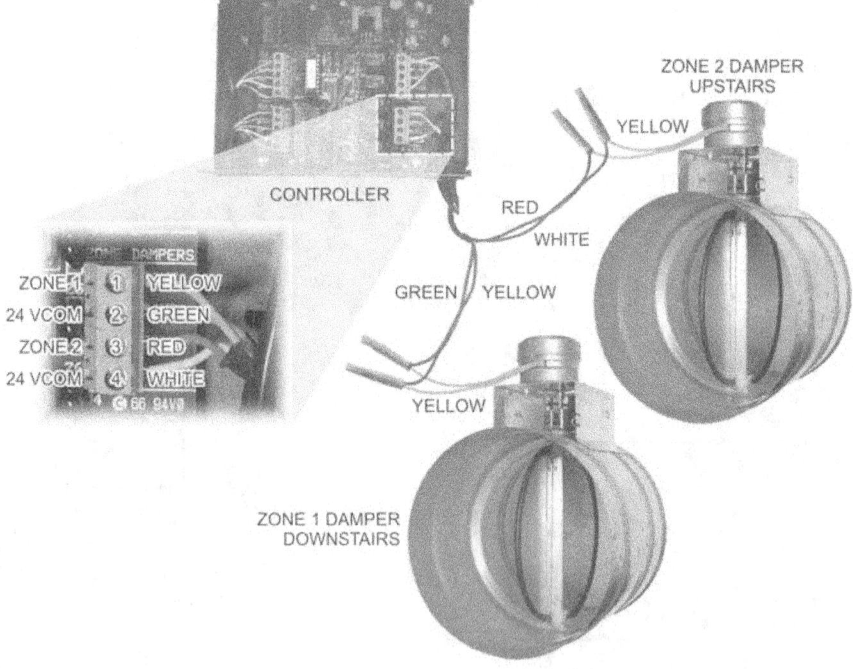

Figure 1-41: Controller and Motorized Damper Wiring

Communication Layer

Communications can be configured in four possible ways with the HVAC communications layer. They are IP based, wireless, serial and proprietary.

1. An IP-based system is ideal because standard off-the shelf computer networking components can be used to control the HVAC system with a LAN-based home computer. Using an IP address also allows the HVAC system to be managed from a Web browser.

2. A wireless connection is the easiest way to configure an upgraded HVAC communications layer. Using an IP address also allows the HVAC system to be managed from a Web browser.

3. A serial interface is a simple way to configure a computer running a HVAC management program to control the HVAC system.

4. A proprietary HVAC control system uses software and control interfaces that are unique to the brand or type of HVAC system. The proprietary system may be more complex to program, however this is not a major problem with most HVAC manufacturers.

LIGHTING TERMINOLOGY AND CONTROL DEVICES

This section addresses the various types of lighting control and management. Applications include outdoor/indoor and centralized and distributed control options

Indoor and Outdoor Lighting Systems

Lighting systems are designed to work under various environmental conditions such as indoor and outdoor applications. Outdoor lighting must be designed to withstand the more severe environmental conditions such as temperature variations, rain and UV radiation. Most indoor lighting uses standard 120 volt power while outdoor landscape lighting often uses low voltage 12 volt systems

Centralized and Distributed Lighting Systems

There are two main types of lighting control systems: **centralized** and **distributed**. With a distributed system the home is wired conventionally, with switches and dimmers in the typical locations. The switches are controlled from a central processor and one or more lighting control keypads or touchscreens.

Know where the load controller is installed for a centralized lighting scheme.

With a centralized lighting control system, the dimming is performed at a central dimming panel. This is usually located in a control room or closet. All the switch wiring from the various lights are run back to the central dimming panel.

Lighting Zones

Zone lighting

Zone lighting is a concept used in lighting management to group lights together for single-switch control. As an example, kitchen lighting may have work area lighting, track lights, and recessed lighting that can optionally be programmed for single-zone, single-switch control. Lighting zones can be configured in a home to automatically control groups of lights called zones. Zone lighting is usually one of the modes available in whole-home lighting systems that are configured to control lights from multiple locations.

In contrast, a 120-volt system runs on the same power that supplies your home and requires the same safety precautions as those observed for normal AC house wiring.

Lighting Scenes

Scene lighting

Scene lighting is one of the features offered by several home automation vendors. Selecting a specific lighting scene is similar to a scenario used by a stage lighting director where a unique lighting condition is programmed in a memory chip or computer and implemented at the touch of a switch.

Pressing a "scene" button on the keypad in the foyer could set the lights not only in that area, but also in the kitchen, family room, and patio. The same command might also lower the temperature and disarm the security system.

Some systems offer a touchscreen display that can adjust the lights to various scene settings. Lighting scenes can be changed for entertaining, evening time, viewing a movie, or for decorative lighting in the daytime hours. The lights can be programmed to automatically shut off after the residents go to bed, saving energy and time. They will similarly turn on to light a path when anyone gets out of bed in the early morning hours.

Occupancy/Motion Sensing

An occupancy or motion sensing system uses IR sensors to determine if anyone is in the room and turns lights off after a period when sensors determine that the room is empty. This type of control system also can be used to activate lights when motion detectors detect the presence of someone entering a protected home security/surveillance area. A sensing light switch is depicted in Figure 1-42.

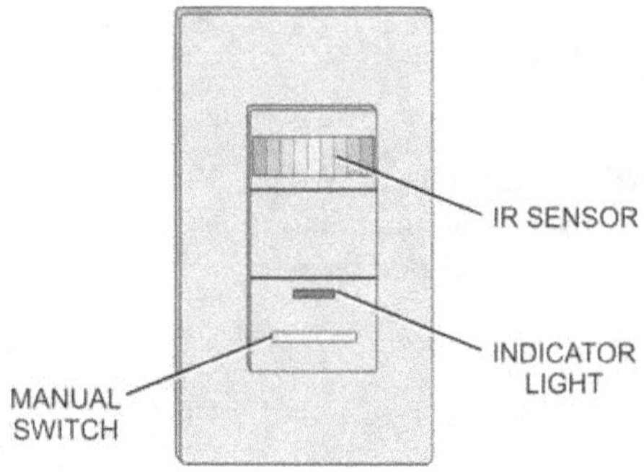

Figure 1-42: Sensing Light Switch

Time and Event Driven Lighting Systems

Control systems are available that can turn off and turn on and lighting systems at dusk and dawn or at a specified clock time settings. Some event driven systems can also be used to control lights when arriving and leaving the home

Window Treatments

Window treatment systems provide the capability to adjust window coverings to control sunlight during daylight hours or close for privacy during nighttime hours, as shown in Figure 1-43. Although they are not normally associated with light control systems, they perform a similar function using control systems such as X-10 and Zigbee.

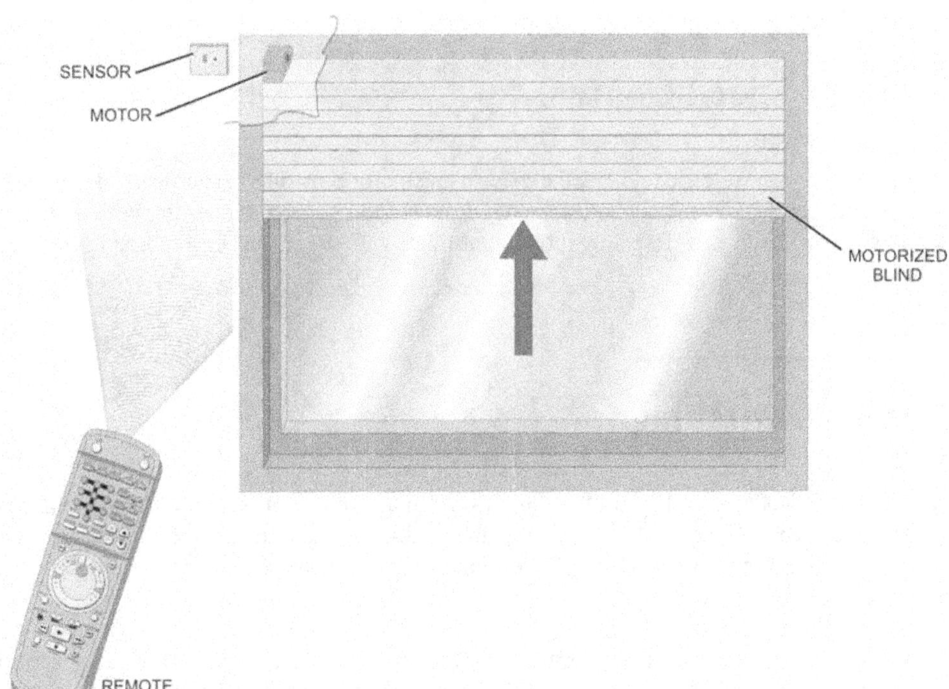

Figure 1-43: Window Treatment System

Energy Management

The National Energy Code Standard ASHRAE/TES 90.1-1999 mandates that all commercial lighting meet strict energy conservation measures. ASHRAE/IES Standard 90.1-2001 features extensive power limits based on square footage of illuminated area or perimeter. Standard 90.1-2001's Section 9 details mandatory provisions related to lighting controls, tandem wiring and exit signs. Standard provisions include automatic shut-off, space controls, exterior lighting controls and additional controls for specialized lighting or applications. Wall switches and ceiling-mounted occupancy sensors are convenient, cost-effective solutions that meet these provisions while providing energy savings and quick payback to end users.

Security Interface

Security lighting can be integrated with automated whole-home lighting systems or can consist of stand-alone lighting fixtures such as motion detector lighting modules located outside the home. Outdoor lighting is used on a nightly basis by many homeowners for safety as well as security and aesthetic reasons.

The two basic types of security lighting systems are photoelectric and infrared. Photoelectric sensors operate by sensing the absence of natural sunlight. Infrared sensors operate security lights by detecting infrared energy emitted by humans moving through an infrared sensor pattern. Motion sensors use infrared technology to turn lights on instantly when motion is detected and off when there is no activity for a desired period of time. Photoelectric sensors discussed earlier in this book measure the amount of daylight present and turn the lights on or off accordingly for added security.

Motor Speed Control

Fans can be controlled to regulate the speed using motor controllers specifically designed for that application. Dimmer switches are not designed to handle the loads imposed on the AC power by a motor, therefore only motor controllers with the correct current rating should be used to control a fan.

Relay/Switching

Light switches are used to control lighting circuits from locations that are convenient to home residents. The types of switches available for residential installations vary according to where they are installed, the type of circuit they control, and the current rating of the load connected to them.

Wall switches are the most popular type of switch for residential light and appliance control. They are designed for easy installation in a wall switch box. The switch contacts are available in various configurations depending on the type of control required.

The **pole** is the number of circuits the switch makes or breaks, and the **throw** is the number of positions to which each pole is switched.

The most common types of poles and contact configurations for switches use the abbreviations and nomenclature given in the next section.

Standard Switch Configurations

The standard types of switch configurations are shown schematically in Figure 1-44. Each configuration is explained in the following list:

- **Single-pole single-throw (SPST)** – This type of switch is used for operating a single light fixture. The incoming hot wire is hooked to one terminal screw, and the outgoing hot wire is connected to the other screw.

- **Single-pole double-throw (SPDT)** – This type of switch is referred to as a *three-way* switch and is used for controlling a light from two locations. It has a double-throw capability and can change the current path between two different wires. A description of a three-way installation for controlling a light fixture from two locations is discussed in the following section.

- **Double-pole single-throw (DPST)** – This switch is another type of on-off control and is used to interrupt a two-wire circuit.

- **Double-pole double-throw (DPDT)** – This type of switch is another changeover type that is normally used to control 240 V, two-phase circuits located at the service entrance panel.

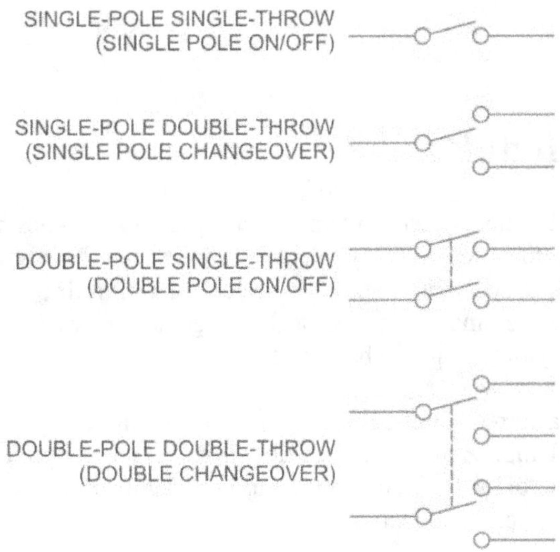

Figure 1-44: Switch Configurations

TEST TIP

Remember the nomenclature and features for each switch configuration, including DPDT, DPST, SPDT, and SPST.

Three-Way Switches

The three-way switch circuit is used frequently in residential installations to operate a light from two locations with one three-way switch located at each position.

Three-way switches look the same as the common single-pole switch, but instead of having only two screws on which to make connections, they have two connections on one side and two on the other side, as shown in Figure 1-45.

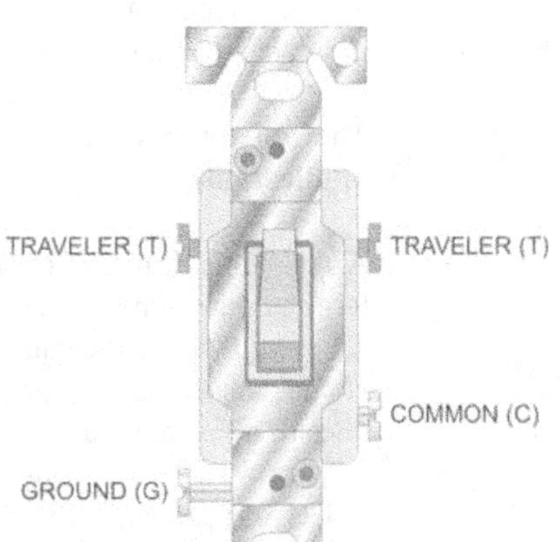

Figure 1-45: Three-Way Switch

One terminal is referred to as the *common* terminal, one terminal is the ground terminal, and the other two are known as **travelers**. This is because the electrical connection goes from the common screw to one of the travelers, depending on the switch position.

This is illustrated in the wiring diagram in Figure 1-46 for controlling a light fixture from two locations. Notice how moving either switch to the opposite position will have full on/off control. As you can observe from the three-way switch diagram, there is no on and off marking for a three-way switch. Moving either switch to the opposite position will change the status of the light from on to off or off to on.

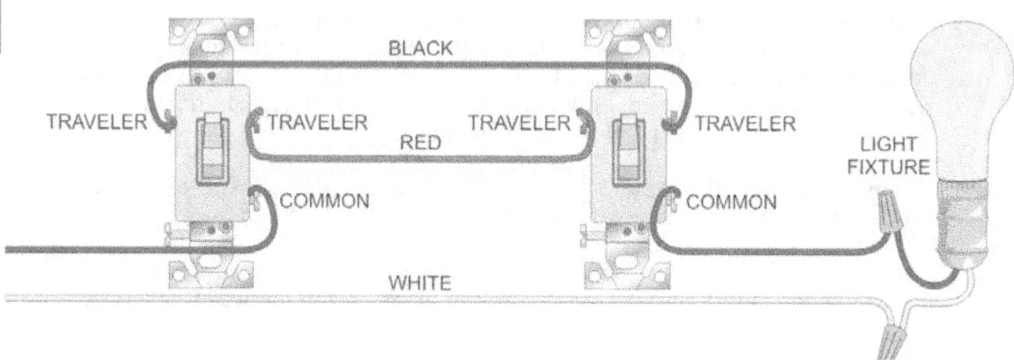

**Figure 1-46:
Three-Way
Switch Wiring**

Dimming Modules

In many residential lighting systems, various lighting areas and zones might not need to be operated at full brightness levels. They might be required to fade in and out and be used at different light levels at different times. A device called a **dimming module** or **dimmer** is required to regulate the amount of electrical voltage sent to each light fixture, thereby allowing the intensity of the light to be varied.

A dimmer contains a current-limiting feature that can vary the voltage distribution between the dimmer and the lamp. Dimmers are available from numerous vendors in many styles to control different types of loads. Older dimmers use a variable resistor to limit the current and resulting voltage available to the lamp.

Modern electronic light dimmers use a variable phase control principle by which the voltage is switched on and off with special components inside the dimmer module.

The components are called **silicon controlled rectifiers (SCRs)** and **thyristers for AC applications (TRIACs)**. A SCR can be thought of as a device that is either on or off.

After an SCR is turned on, it turns off only when no current is flowing through it. It has three terminals and schematically looks just like a diode with an extra wire coming off the anode. This extra wire is the gate, and very little current is required to switch on the SCR. An SCR conducts current only one way, so if the current changes polarity, it turns off because it doesn't allow any current to flow.

A TRIAC schematically resembles two SCRs placed back to back with only one gate coming from one anode. This enables AC operation, but every time the current changes polarity, the TRIAC turns off unless the gate is held high.

Variable phase control, mentioned earlier, refers to allowing only portions of the AC cycle to go through the load. It is usually performed with a TRIAC or two back-to-back SCRs. AC current sources have a zero-crossing two times every cycle. This happens 120 times per second for 60 hertz (Hz) AC service.

The TRIAC turns off 120 times per second. By varying the turn-on period, the amount of power applied to a lighting circuit can be varied. This produces the waveform shown in Figure 1-47.

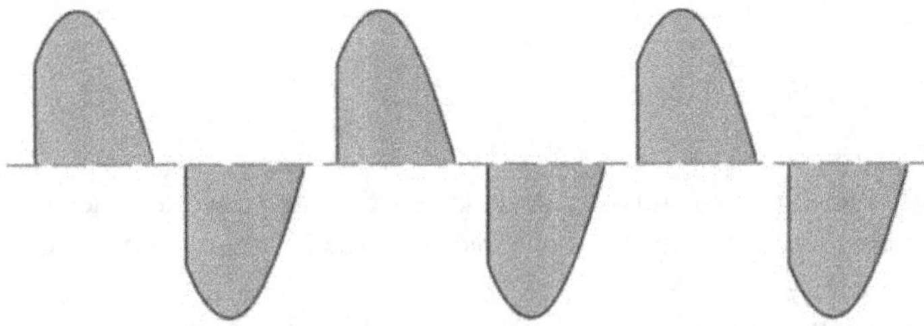

**Figure 1-47:
TRIAC Waveform**

To deliver only half the power, the TRIAC is fired between each zero-crossing, producing the AC waveform shown in Figure 1-48.

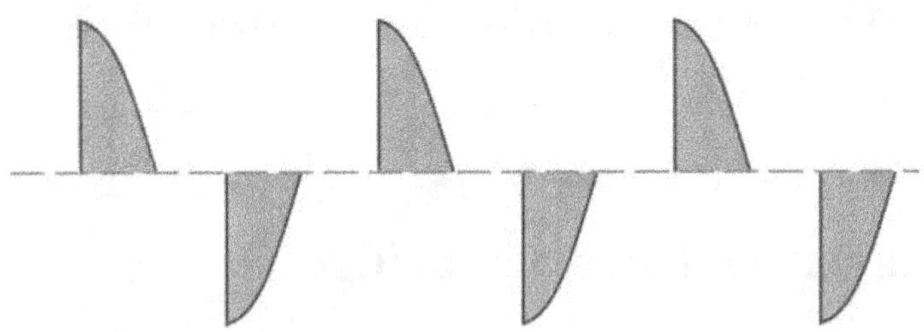

**Figure 1-48:
TRIAC Half-Power
Waveform**

This type of module is normally used for light dimming because lights are resistive loads and do not flicker even if they only get voltage spikes, as illustrated in Figure 1-47.

TRIAC dimmers are designed to work exclusively with incandescent lighting. Fluorescent lighting requires a special ballast designed for dimming and a special dimmer (one that indicates that it is for a dimming fluorescent fixture). Ordinary fluorescent fixtures are not designed for dimming.

Additional information on fluorescent lighting is covered earlier in this book under the heading "Fluorescent Lighting Fixtures".

Photoelectric Switches

Photoelectric switches are a special type of switch used for controlling lights depending on the light intensity (lux value). This type of switch contains a special sensor that operates when the light intensity or darkness reaches a certain value. Photoelectric switches are typically used for outdoor locations for landscape and security lighting.

Most photoelectric switches are integrated in a single fixture that includes the sensor and lamp fixture. Most have adjustments to turn lights off after a number of hours where it is not necessary to have lighting from dusk to dawn. They also have a time delay of approximately 15 seconds to prevent the fixture from turning on or off in cases of short-duration intensity changes.

Photodiodes are often used as skylight sensors to precisely monitor either work or ambient light levels.

The measured light levels are converted to analog signals that are sent to an automated system controller that manages energy use for the entire residence. The sensors provide the information required to control area lighting by switching banks of lights on and off or by operating electronic dimming ballasts for fluorescent fixtures.

Clock Switches

Clock switches turn lights on and off at given intervals. They are often used to control lights when the members of the family are away and want the home to appear occupied.

Clock switches can be inserted into a standard wall outlet. The clock obtains power from the AC receptacle. Lamps or appliances can be plugged into the clock timer switch. Programmable tabs can be arranged on the dial to turn the light on or off at any time.

Clock timers are rated for the amount of power, current, and voltage that they can deliver to the load connected to the outlet.

Communication Interface/Bridge

A communications interface device called a **bridge** is used to couple the two-phase AC distribution system in home power wiring.

The following are examples of problems that may appear when installing and operating X-10 receiver and transmitter modules:

- Some receiver modules work properly, but others work intermittently.

- Some transmitters and receivers may communicate with each other, while others only a few feet apart may not work at all.

- Problems disappear when the electric stove is operating.

- Lamps or appliances operate when they are not activated.

Many of the problems listed above are caused by the isolation of the two phases of the 120-volt power lines. Homes in North America are wired for two-phase electric power distribution.

The 240-volt service is split between each leg on the 240-volt service, and a neutral wire that yields two 120-volt phases for home lighting and appliances. X-10 receiver modules installed on a separate phase from the controller module may only work sporadically because the only coupling between the two phases is through the pole transformer outside the home. The transformer blocks the X-10 signals from reaching the opposite phase.

A bridge (also called a phase coupler) can be installed to bypass the transformer, providing a low-inductance path between the two phases. This usually solves the problem of intermittent operation, or when some of the modules do not work at all.

An X-10 **signal bridge/repeater** (also known as a **phase coupler/repeater**), shown in Figure 1-49 provides coupling between the two 120-volt phases, and also regenerates the signal to maintain adequate signal levels on each leg of the 120-volt circuits. The bridge/repeater model shown functions as the signal bridge, but it also regenerates the X-10 signal and transmits the signal out both legs of the 120-volt circuits. Provided there is at least 100 mV of incoming signal, this unit will boost the outgoing signal to over 5 volts.

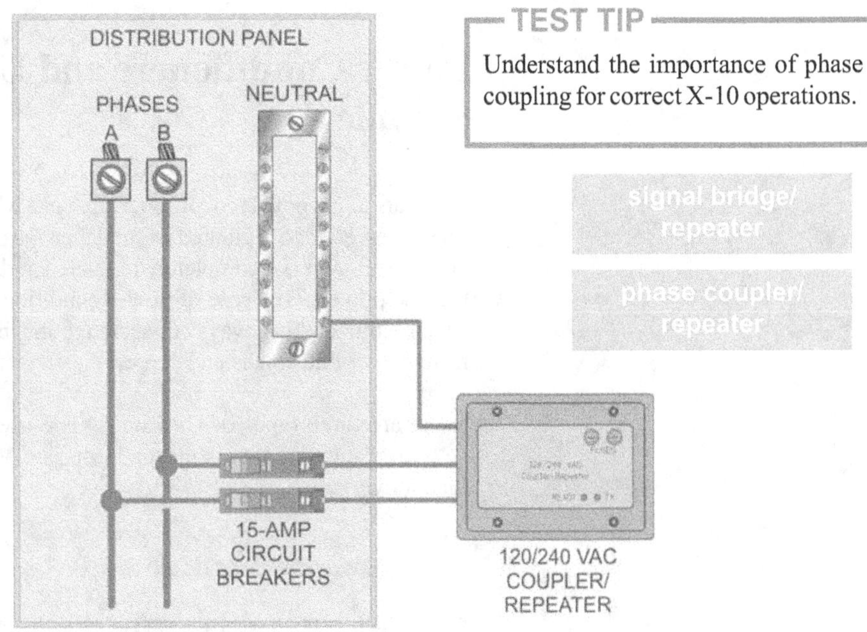

Figure 1-49: Signal Bridge/Repeater

COMPONENT POWER PROTECTION DEVICES

Home computer systems and other electronic components that use AC commercial power are vulnerable to spikes and surges on the input voltage. These irregularities are due to transient noise, lightning and poor regulation on the power lines, Power conditioners, surge suppressors and **uninterruptible power supplies (UPS)** are available in different configurations to guard against damage to home electronic systems.

Surge Suppression

A device that protects computers and home theater systems from irregular input line voltage, spikes and surges is called a **surge suppressor**. A surge suppressor attempts to regulate the voltage supplied to an electric device by either blocking or by shorting to ground voltages above a safe threshold. A power strip is sometimes incorrectly called a surge protector, although some power strips have surge protection built-in and are clearly labeled.

Surge suppressors are available that safeguard a complete home. The most common design of whole-house surge suppressor is mounted in the breaker panel. Another design is used as a base under the electric meter.

A point protection type surge suppressor is designed to protect equipment within the home such as computers, audio and video equipment and modems.

Power Conditioners and Uninterruptible Power Supplies

There are some power conditioners that use what is called double conversion, where the AC line is converted to regulated DC and then the output is another conversion back to AC. What you end up with is a completely regulated and shaped sine wave output that is isolated from the line power. This type of power conditioner is recommended over the surge protectors because the load is always connected to the backup power source and is therefore isolated from power line spikes and surges.

There are three types of UPS units. The use and the degree of protection required, will determine the type that is most appropriate. The three types are Offline, Online, and Line Interactive.

Standby UPS (Offline)

Two configurations of UPS power protection systems are the offline and online types. As shown in Figure 1-50, the AC line in the standby UPS offline unit is the primary power source. When the line sensor detects a loss of AC power, it switches the battery to provide power to the inverter, cutting of the main power line-thus allowing the battery to supply the required power. The transfer time should not take more than 4 milliseconds, as this is the tolerance limit of a typical computer power unit.

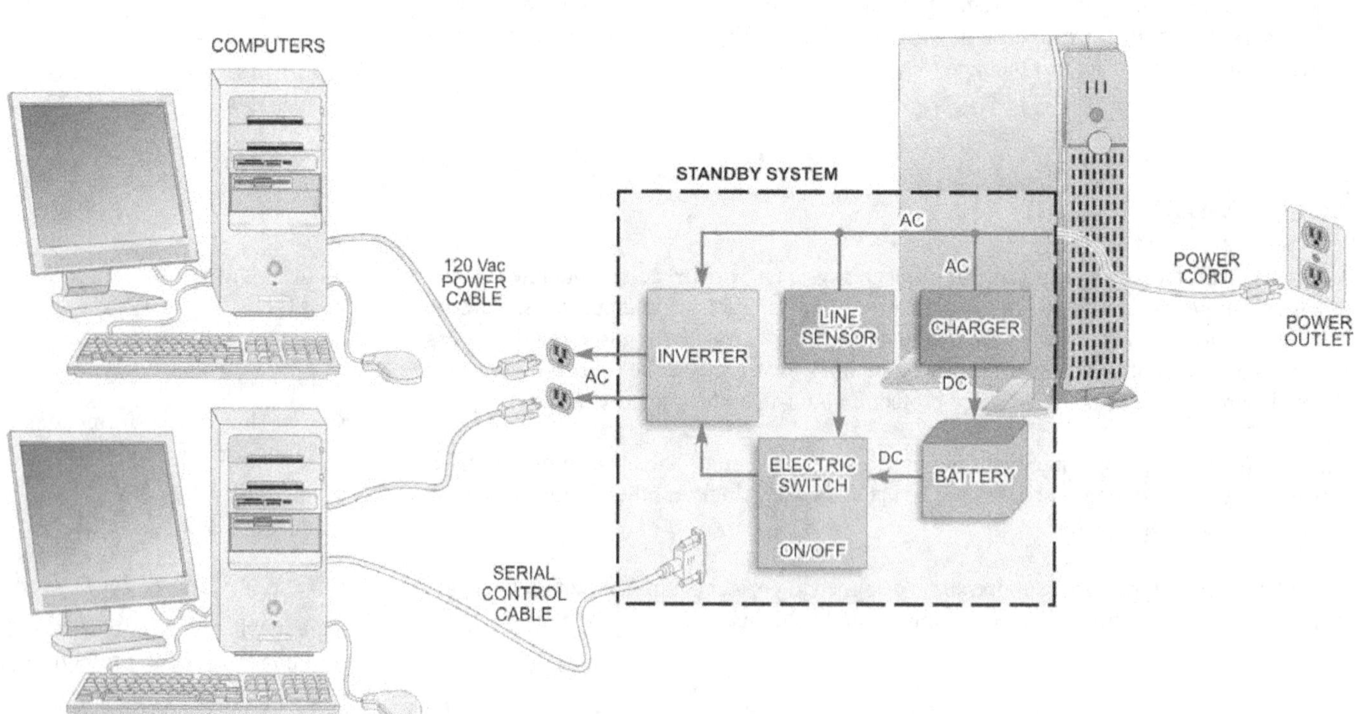

Figure 1-50: Standby UPS

Online UPS

The UPS online unit operates with the inverter as the primary power source at all times; therefore there is no transfer time in case of power failure. As shown in Figure 1-51, the AC power source supplies power to the charger and the charger in-turn provides power to the inverter. Since there is no changeover during an AC power interruption power is continuously supplied to the output load without the requirement for a changeover as is the case with the offline UPS unit. Since the battery acts as the primary power source to the system, an online UPS battery charger has to be powerful enough to generate enough power to compensate for the battery's power drain.

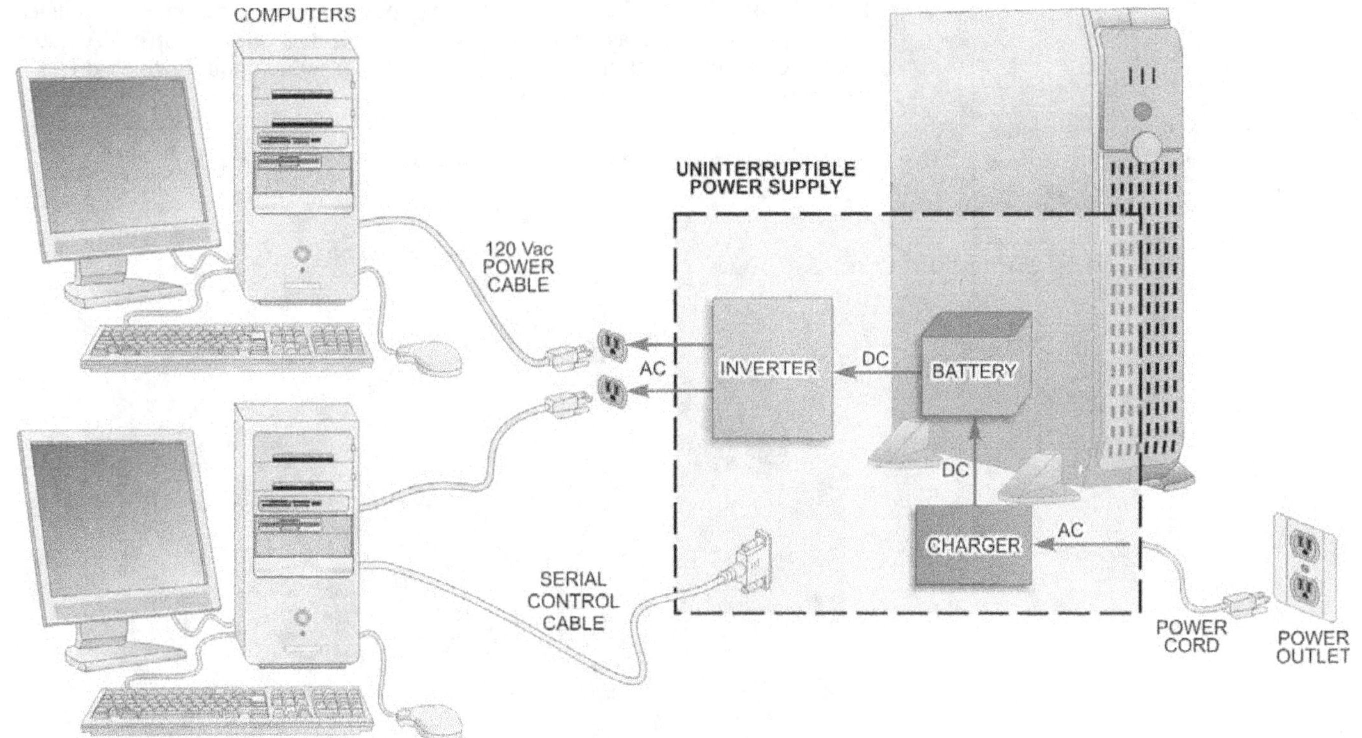

Figure 1-51:
On-line UPS

Line-Interactive UPS

With the line interactive UPS, the AC line power is the primary source and is being constantly filtered. In case of a brownout condition the automatic voltage regulation circuitry is activated to correct the problem. The battery and inverter circuitry supplies the backup power in case of a total blackout.

Residential Backup Power Options

Designed to be permanently installed outside the home, an emergency backup power generator is fueled by gasoline, diesel, natural gas, or propane. It can be activated automatically or manually to supply power after a utility power interruption or outage.

Residential backup power systems can be sized and wired for a home in one of two power distribution designs described as follows:

- **Whole-house power distribution** – This type of power generator is sized for powering all devices within the home. It is normally driven by a gasoline engine.

- **Essential power distribution** – This type of power system is designed for isolated essential devices such as a security system, a freezer, a refrigerator, HVAC, and limited emergency lighting. It is powered by a gasoline engine, which drives a generator.

An example of an essential power backup system is shown in Figure 1-52.

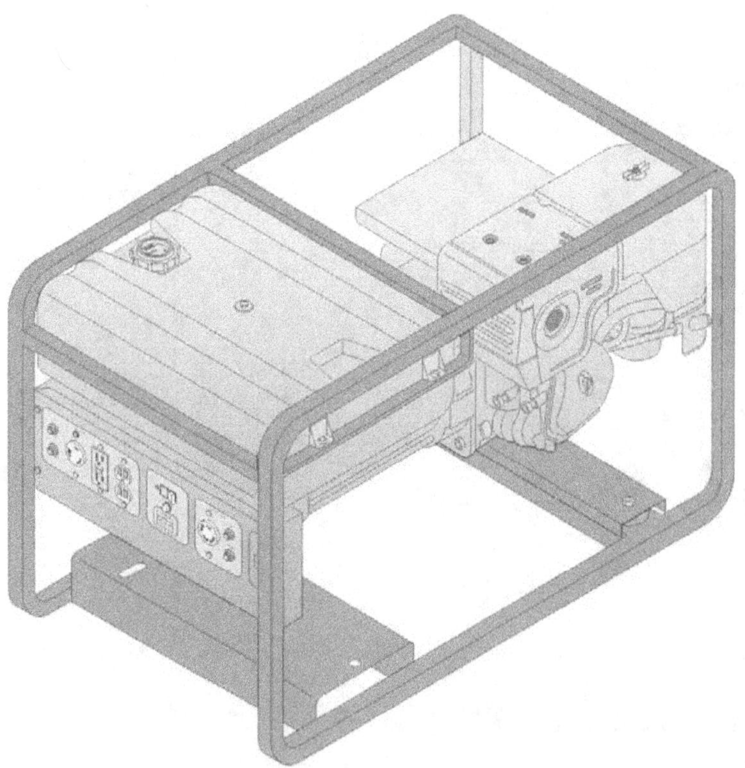

Figure 1-52: Essential Backup Power System

Residential power backup systems use a transfer switching device to transfer the electrical load to the generator and disconnect the commercial power source. Automatic switches provide an automatic signal to start the generator and switch over the load to the generator when commercial power has failed. When commercial power is restored, the switch automatically returns the residential load to the commercial power source and shuts down the generator. Most residential backup systems have voltage regulators to maintain a constant voltage under varying load conditions.

KEY POINTS REVIEW

Review the following key points before proceeding to the Review Questions and Exam Questions sections to make sure you are comfortable with each point. After completing the review, answer the Review Questions that follow to verify your knowledge of the material covered in this book.

- A residential controller is a device that integrates all of the automated home features into a coherent system.

- Electronic key fobs are used for remote keyless entry systems on motor vehicles.

- Protocols serve to assure all devices can communicate with each other regardless of the manufacturer/supplier.

- The most widely used I/O interface scheme in current PCs is the universal serial bus (USB).

- All computers, routers, or network-enabled devices that access the public Internet today must be assigned an Internet Protocol (IP) address.

- IrDA is an international organization that creates and promotes interoperable infrared data interconnection standards for wireless networks.

- For HVAC installations, zoned system designs should be considered when the residence is multilevel, or a large single story structure.

- Clock switches are used to turn lights on and off at given intervals.

- The standard ASCII code uses seven-digit binary numbers. The code can represent 128 different characters, since there are 128 different possible combinations of seven 0's and 1's.

- Universal powerline bus (UPB) is a home control system that uses the existing power wiring to transmit control information between UPB devices.

- Remote controls are used to issue commands from a distance to televisions or other consumer electronic equipment.

- Modern HVAC systems are generally equipped with smart electronic control systems. Smart thermostats connect into the HVAC system via the existing residential wiring and communicate with the various control modules through three-conductor control cables.

- The USB management software dynamically tracks what devices are attached to the bus and where they are. This process of identifying and numbering bus devices is known as bus enumerating.

- X-10 is a popular standard Power Line Carrier (PLC) signaling and home control protocol language that enables signals to be transmitted anywhere in the home over existing 120-volt electrical lines.

- WiFi (Wireless-Fidelity) is a logo from the Wi-Fi Alliance that certifies network devices comply with the IEEE 802.11 wireless Ethernet standards.

The following questions test your knowledge of the material presented in this book.

1. What standard is used to monitor alarm panel status for conditions such as open door, temperature limits, and power failure or intrusion alarms?

2. What specification for a wireless communication protocol uses small, low-power digital radios based on the IEEE 802.15.4 standard?

3. What communication technique uses the existing power wiring to carry information?

4. What type of thermostat provides greater efficiency and lower operating expense by providing multiple levels of heating and cooling capacity?

5. What are the two main types of residential lighting control systems?

6. What type of switch is referred to as a *three-way* switch and is used for controlling a light from two locations?

7. What are the three types of uninterruptible power supplies (UPS) used for the protection of home computer and electronic components?

8. What type of thermostats are being phased out and replaced by programmable thermostats?

9. What speeds and range of operations are supported by Bluetooth?

10. What is the maximum segment length for an IEEE-1394 connection?

11. What are "keyfobs"?

12. What are the two Electronic Industries Alliance (EIA) standards frequently included as a built-in feature with home controllers for interfacing with external devices?

13. What are the basic features offered by a PDA?

14. What is the primary difference between a zoned and non-zoned HVAC installation?

15. What are the main advantages offered by the FireWire interface when compared to other input/output systems?

1. What standard wireless technology is designed to connect one device to another with a short-range radio link?
   a. FireWire
   b. EIA 232
   c. Internet protocol
   d. Bluetooth

2. What type of switch is used for controlling lights depending on the light intensity (lux value)?
   a. Photoelectric
   b. Clock
   c. DPDT
   d. 3-way

3. What is the purpose of signal bridge/repeater used with X-10 control systems?
   a. Noise reduction
   b. Provides coupling between the two 120-volt phases
   c. Provides protection from lightning strikes
   d. Elimination of radio frequency interference (RFI)

4. What wireless standard for networking is now "built-in "with most hand-held devices and laptop computers?
   a. Zigbee
   b. 802.3
   c. Bluetooth
   d. 802.11

5. Which of the following is an RF-based two-way mesh networking protocol that allows for the control of thermostats, garage door openers, light switches, and residential appliances?
   a. WiFi
   b. Z-wave
   c. IP
   d. EIA232

6. What is the effective data rate provided by the X-10 protocol?
   a. 100 Mbps
   b. 18.62 bps
   c. 10 Mbps
   d. 182.6 kbps

7. What is the purpose of dampers used with HVAC systems?
   a. Dampers are typically used in a zoning application to direct air to the areas that need it most.
   b. Dampers are used with manual thermostats to close off unoccupied areas.
   c. Dampers are only used when heating a zoned area.
   d. Dampers are used to suppress the fire hazard in HVAC installations.

8. What are the two basic types of security lighting systems?
   a. Imaging and photoelectric
   b. Sound and motion detectors
   c. Electronic and manual
   d. Photoelectric and infrared

9. What type of lighting control systems uses programmable tabs to turn the
light on or off at any time?
a. Clock switches
b. Dimming controls
c. DPDT switches
d. Photoelectric switches

10. What type of power protection device operates with the inverter as the
primary power source supplying the load and therefore has no transfer time
in case of power failure?
a. Surge suppressor
b. Standby UPS
c. On-line UPS
d. Power inverter

LAB
PROCEDURES

A Residential X-10 Power and Control System

OBJECTIVES

1. Operate a simple residential power and control system.
2. Program the operation of a basic power and control system.
3. Test basic power control program operations.

Power and Control Systems

RESOURCES

1. Marcraft Residential Power and Control Experiment Panel and Frame
2. X-10 Powerhouse ActiveHome Owner's Manual
3. Computer system with Windows XP Professional
4. X-10 HC60RX, RCA ActiveHome Kit
5. X10, RR501 Transceiver Molule
6. 2 - X10, LM465 or PLM03 Lamp Modules
7. Floppy disk, 3.5" (work disk)

TOOLS

1. Flat-tip screwdriver

DISCUSSION

Once the installing technician has carefully checked the physical integrity of the various components of the residential power and control system, the next logical step is to ensure that all electrical components operate properly. Then, basic programming can be performed and the computer software program can be utilized to operate the system according to the information given to it. If this were an actual field operation, the installing technician would carefully consult with the customer at this point to determine his or her needs and work requirements, including those for other family members. This means that an actual residential power and control installation would eventually include solutions for those control functions that may be unique to the client's personal situation.

In the classroom, however, the best approach is to program only a basic set of functions into the residential power and control system. This will provide the opportunity to become familiar with its overall operational strategy.

PROCEDURES

Operating a Simple Residential Power and Control System

Keep in mind that the computer interface module is the essential component required for all of the programming activities. It should be connected to the computer through the COM 2 serial port. No other wiring is needed to begin programming the system.

1. Before doing anything else, carefully check the Marcraft panel and the residential power and control computer system.

The control box on the panel should be closed and the hardwire connection between the panel's control interface module and the computer's serial port should exist. While all power switches (on the panel and the computer) should be OFF, both the panel and the computer should be plugged into an active ac power source.

NOTE: All manual nightlight toggle switches should be in their ON (left) positions, as shown in Figure 1-1.

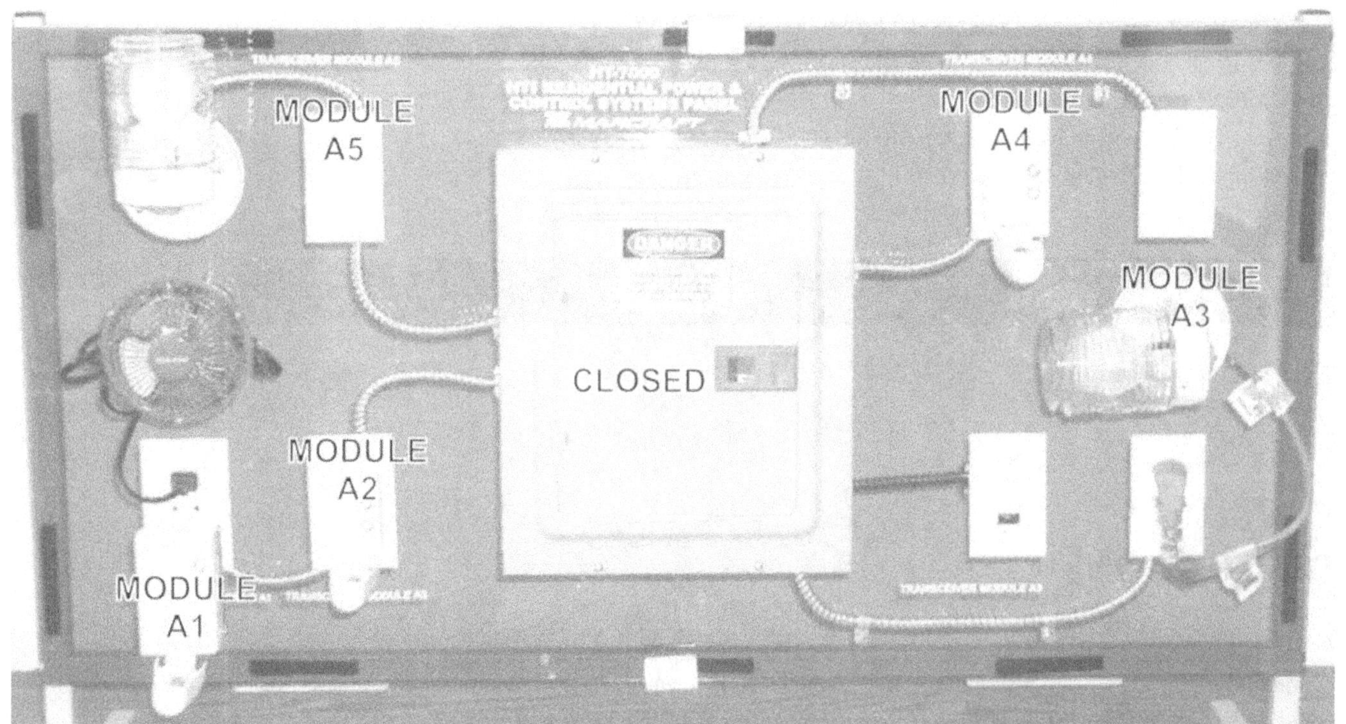

Figure 1-1: The Marcraft Panel

2. Configure the Marcraft panel and computer as described above.

3. Apply power to the Marcraft panel by moving its power toggle switch (lower-right) to its **ON** (left) position, as shown in Figure 1-2.

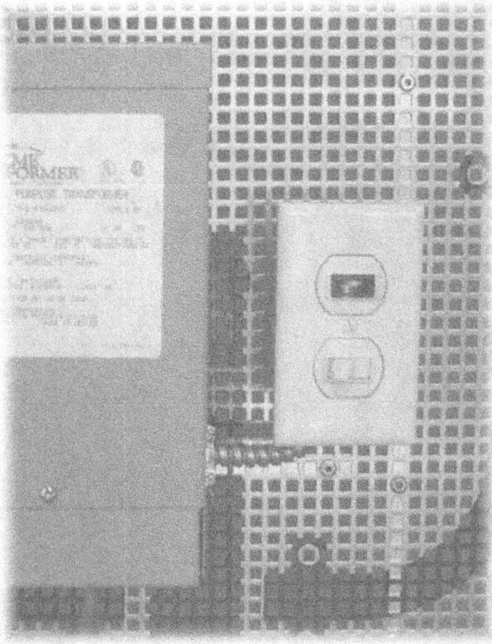

Figure 1-2:
The Marcraft
Panel Power
Toggle Switch

The red indicator light should illuminate to verify that power has been applied to the transformer, because it's extremely important to know precisely when the transformer is operating. Without this type of visual confirmation it could be mistakenly assumed that the Marcraft panel is not energized.

4. Starting with the *A1* transceiver module, verify that power is available to its circuit branch by pressing the ON/OFF button oin the upper right of the transceiver.

The nightlight plugged into the bottom outlet of the transceiver should quickly turn ON.

5. Locate the key chain remote control unit and press the top right *OFF* button, as shown in Figure 1-3.

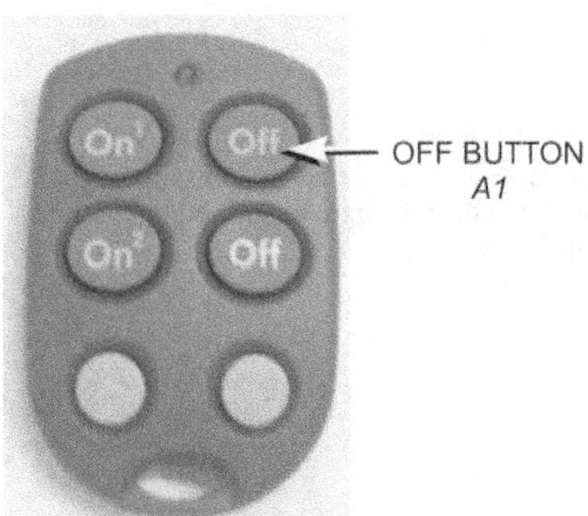

Figure 1-3:
Turning Module A1
OFF Using Key Chain
Remote Control Unit

The nightlight plugged into the bottom outlet of the transceiver should quickly turn OFF, as shown in Figure 1-4.

6. On the key chain remote, press the top left **ON** button to turn the light controlled by module *A1* back ON.

7. Locate the 6-in-1 remote control unit, and press the **X-10** button, as shown in Figure 1-5, to ensure that the remote is ready to transmit X-10 data.

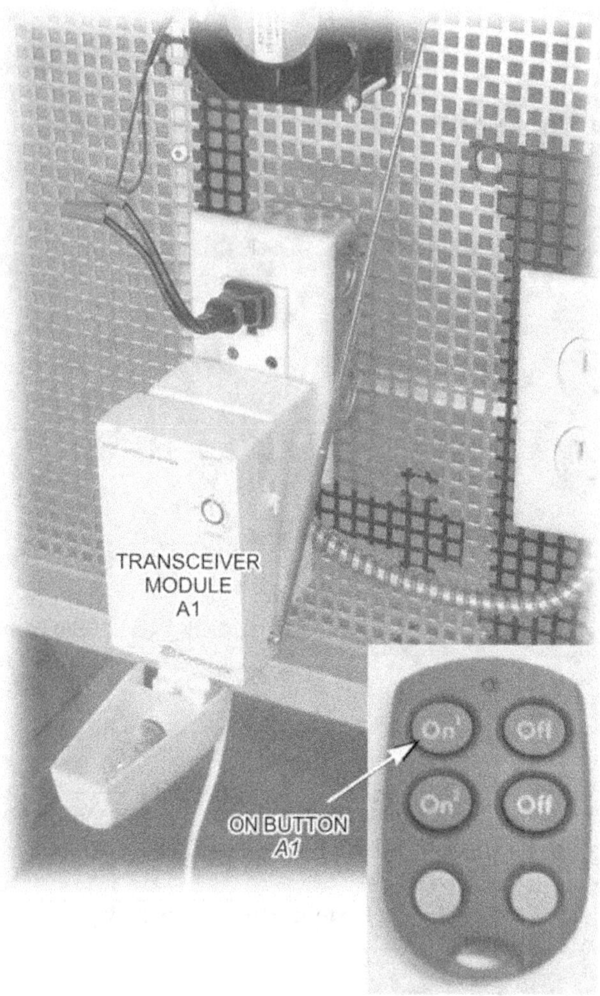

**Figure 1-4:
Turning the Light
at A1 ON**

**Figure 1-5: Pressing the X-10 Button
on the Remote Control Unit**

8. On the 6-in-1 remote control unit, press the **M** (Mute) button to turn the light controlled by module *A1* back OFF.

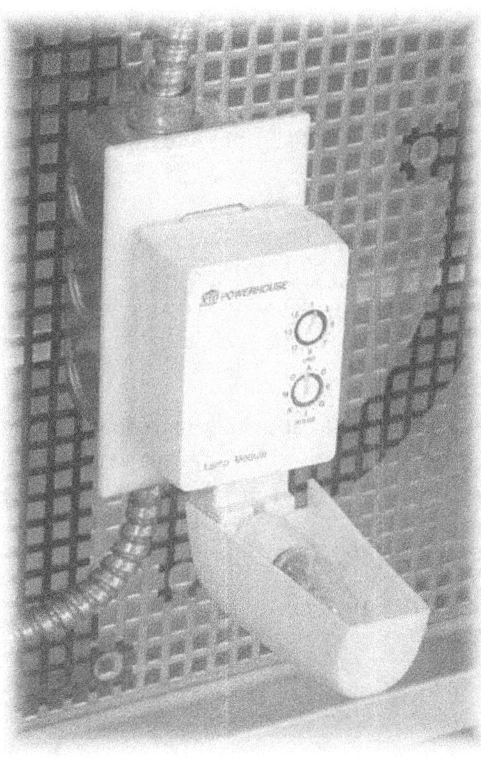

Figure 1-6: Turning the Light at A2 ON

9. At module *A2*, manually move the nightlight's toggle switch first **OFF**, and then **ON**. The nightlight shown in Figure 1-6 should illuminate fully.

10. Locate the key chain remote control unit and press the bottom right **OFF** button. The nightlight should fully extinguish.

11. Using the 6-in-1 remote control, test the operation of the 240 Vac light unit at module *A3* shown in Figure 1-7 by pressing the **3** button, and then pressing the **CHANNEL+** rocker button.

Figure 1-7: The 240 Vac light at Module A3

12. Now press the **CHANNEL–** rocker to turn the 240 Vac light unit OFF.

13. See if the 240 Vac light unit will activate by pressing the **3** button, and then pressing the **VOLUME–** rocker button.

14. If the 240 Vac light unit does activate, press the **CHANNEL–** rocker to turn it OFF.

15. Illuminate the nightlight plugged into module *A4*, as shown in Figure 1-8, by manually moving its toggle switch first **OFF**, and then **ON**.

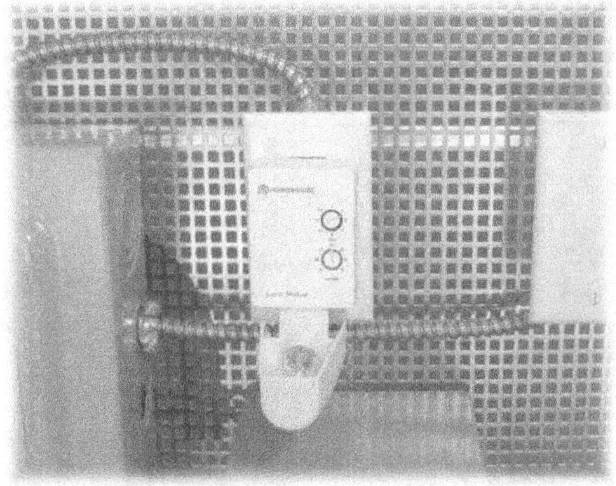

Figure 1-8: Turning the Light at A4 ON

16. On the 6-in-1 remote control unit, press the number button **4**, which corresponds to the unit code for module *A4*.

The default house code for the 6-in-1 remote control unit is *A*. If necessary, the house code can be changed to any letter between *A* and *P*. For purposes of the residential power and control system lab procedures, the house code used will always be *A*.

17. Now, press the **CHANNEL–** rocker button to extinguish the nightlight plugged into module *A4*.

18. Press the **CHANNEL+** rocker button to illuminate the nightlight once again. Then, press the **CHANNEL–** rocker button.

Table 1-1 specifies the X-10 functions available on the 6-in-1 remote control. Keep in mind that appliance modules cannot be dimmed or brightened. They simply react to the codes they receive by alternating to their opposite states. For example, if an appliance module is OFF and receives either a *VOLUME+* or *VOLUME–* code, the module will merely turn ON.

Table 1-1: 6-in-1 Remote Control X-10 Functions

Function	Button
On	CHANNEL+
Off	CHANNEL–
Bright	VOLUME+
Dim	VOLUME–
All Lamps On	POWER
All Modules Off	MUTE

19. Test the wall switch at module *A5* shown in Figure 1-9 by pressing its push button once.

The 40-watt lamp in the ceiling fixture should illuminate fully.

**Figure 1-9:
The Wall Switch
at Module A5**

20. Then, slide the module's "kill" switch (located below its push button) to the left.

> ## CAUTION
>
> Avoid looking directly at the 40-watt lamp! Even though it is only 40 watts, it is still strong enough to cause a spot in your vision for several minutes if you stare at it.

The ceiling light at *A5* goes out.

21. With the kill switch still in its disable (left) position, try to activate the ceiling light at *A5* by pressing the wall switch's pushbutton repeatedly.

With the kill switch in its disable position, the pushbutton cannot activate the ceiling light.

22. Slide the module's "kill" switch to the right.

23. Now press the wall switch's pushbutton at *A5* to activate the ceiling light.

24. Press the **5** button on the 6-in-1 remote control and press the **CHANNEL**– rocker to turn the ceiling light OFF.

25. Then, press the **CHANNEL**+ rocker button to turn the ceiling light back ON.

26. Now press the **VOLUME-** code button to dim the ceiling light to some level below maximum.

27. Press the remote's **VOLUME**+ code button to maximize the ceiling light's level and then press the **CHANNEL**– rocker once more to turn the ceiling light OFF.

28. Using the 6-in-1 remote control, test the operation of the 120 Vac fan unit at module *A6* shown in Figure 1-10 by pressing the **6** button, and then pressing the **CHANNEL**+ rocker button.

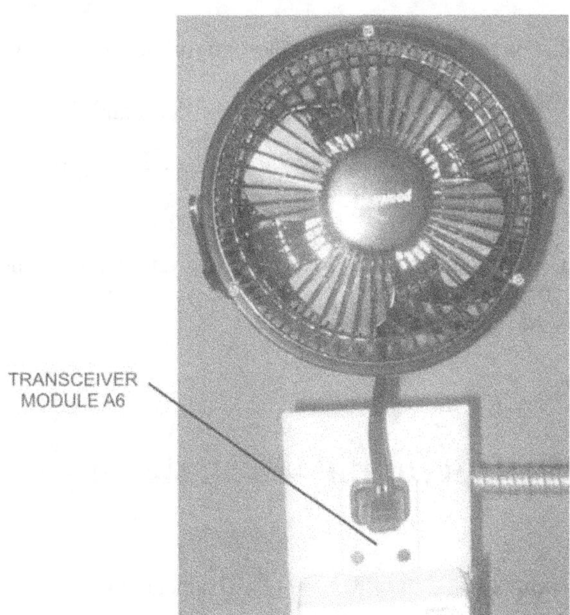

**Figure 1-10:
The 120 Vac Fan Unit
at Module A6**

29. Now press the **CHANNEL**– rocker to turn the 120 Vac fan unit OFF.

30. See if the 120 Vac fan unit will activate by pressing the **6** button, and then pressing the **VOLUME**– rocker button.

31. If the 120 Vac fan unit does activate, press the **CHANNEL**– rocker to turn it OFF.

32. Press the **MUTE** button on the 6-in-1 remote control to ensure that all modules are turned OFF.

33. Observe the results at the panel by pressing the **POWER** button on the 6-in-1 remote.

All modules designed to control lamps should have reacted to the "All Lamps On" command signal by turning their associated lamps ON.

34. Again press the **MUTE** button on the 6-in-1 remote control to turn all modules OFF.

35. Press the **1** button on the 6-in-1 remote control and then press the **CHANNEL**+ rocker.

The nightlight plugged into the transceiver module should illuminate, because its outlet is assigned to the *A1* code. In fact, any device plugged into the transceiver's outlet would have been activated on receipt of an *A1* code. Although the transceiver activated the nightlight plugged into it, it does not operate in the same way a lamp module does. It disregards commands that are targeted strictly to lamp modules, including any dimming instructions.

36. Turn all modules OFF again by pressing the **MUTE** button.

37. Using a flat-tip screwdriver, change the house code on the lamp module to the right of the control box from *A4* to **C4**.

38. Press the **POWER** button on the 6-in-1 remote control and observe the results at the panel.

39. Press the **MUTE** button to turn all activated modules OFF.

40. Change the house code on the lamp module to the right of the control box back to **A4**.

41. Again press the **POWER** button on the 6-in-1 remote control and observe the results.

42. Press the **MUTE** button once more to turn all activated modules OFF.

Once you have tried the various modules on the Marcraft panel and the two remote controllers, you can begin programming the residential power and control system. The programming is accomplished through using the X-10 software installed on the computer system.

Programming the Operation of a Basic Power and Control System

1. Verify that both the computer and the HT 7000 (X10) panel are **Off**.

2. Remove the transceiver module (RR501) from the serial interface module (HC60R); located in the lower left receptacle (**A1**) of the Marcraft panel.

3. Install two (2) fresh AAA batteries in the serial interface module and replace the access cover.

NOTE: The electrical receptacle for A1 & A6 is a dual-purpose outlet. The lower half (used for the Serial Interface and Transceiver Modules) and is normally open; allowing electricity to flow through it normally. However, the upper half is a controlled outlet, utilizing the two dials to set the house code and location. Ensure these two dials are set correctly; the left, red dial on "A" and the right, black dial set to "6".

4. Reinstall the transceiver module back into the outlet of the serial interface module.

5. Using the Serial/RJ11 conversion cable supplied with the X10 panel, plug the RJ11 end into the serial interface module.

6. Next, plug the serial end of the cable into the chosen COM port of your computer.

NOTE: If you have more than one COM (Serial) port on your computer, choose the one Not being used by another program. If only one COM port is available some reconfiguring may need to be done and will be explained later in this section.

7. Turn On the computer being used for this section of the lab.

8. While waiting for the computer system to boot-up, turn the X10 panel's power **On** and make sure no lights or appliances are 'running' at this time.

NOTE: If you have previously installed the HAI Web Link II software on this machine continue to Step 9. If the software has Not yet been installed on your system, proceed to Step 13.

9. Once the system has finished boot-up, press the **Ctrl+Alt+Del** keys simultaneously and open the Task Manager.

10. Click on the **Processes** tab and search for a file called *"HAICommSrv.EXE"*.

11. If this process is found, end the process before proceeding with the lab.

12. When completed, exit the *Task Manager* and continue.

13. If not previously done, install the X10, ActiveHome software using the suggested default settings; **except as noted below**. If the ActiveHome software has already been installed proceed to Step 14.

NOTE: By default, the ActiveHome software chooses COM 2 to use during it's communication with the serial interface module. During installation, there is an opportunity to change this from the default location to on of the user's choosing. It is suggested that COM 1 be selected during installation for ease of component operation during this lab.

14. Launch the ActiveHome software program and look for its opening screen, similar to that shown in Figure 1-11.

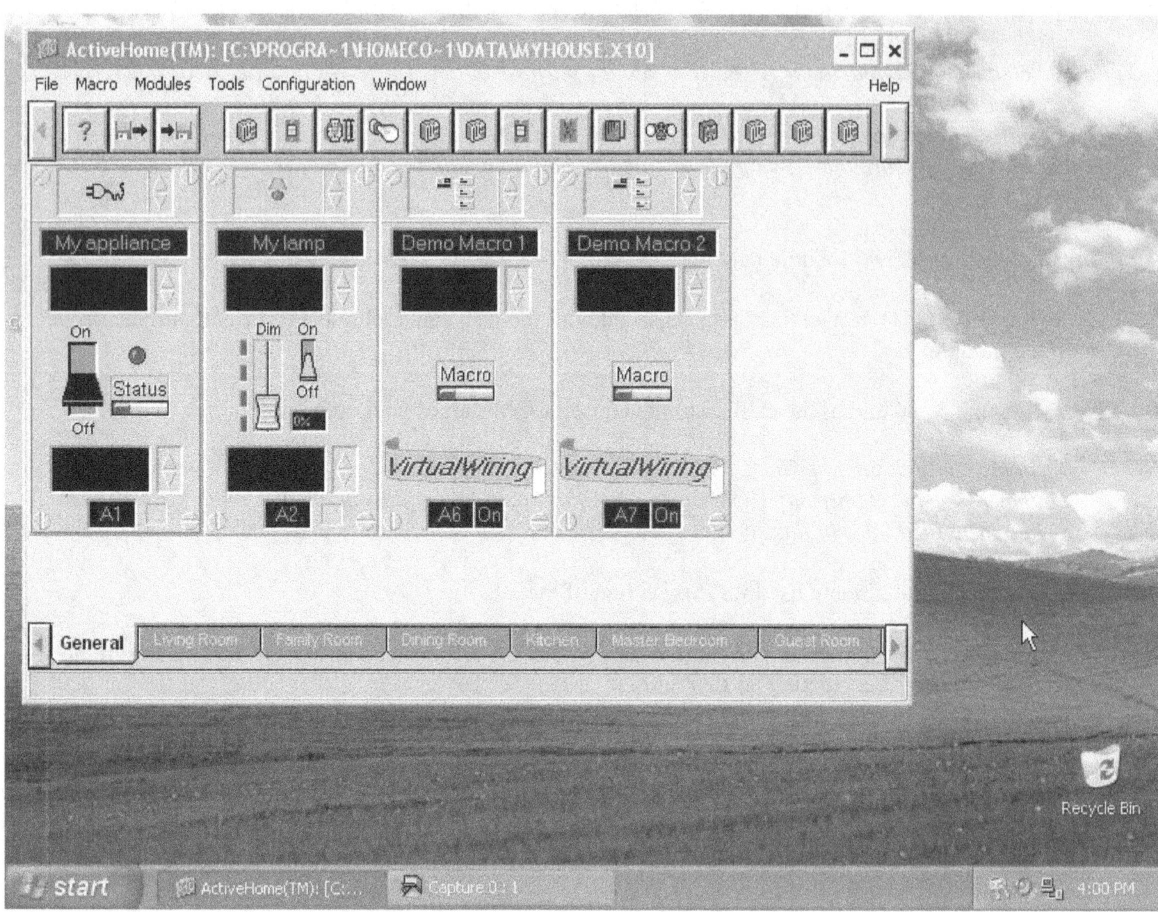

Figure 1-11: Opening Screen for ActiveHome Software Program

Above is a demo program, MYHOUSE.X-10, that automatically launches when ActiveHome is first started. Notice that it displays a graphical representation of a transceiver module (titled "My appliance") and a lamp module (titled "My lamp").

Figure 1-12: Work Area of a Cleared Screen

15. Click on the **MAXIMIZE** button in the upper right corner of the program window to fill the available screen area with the program.

16. Press the **F2** function key, or select the *New* command from the *File* menu listing.

The work area of the screen is cleared, as shown in Figure 1-12, to permit the creation of a new arrangement of modules.

17. Click on the icon in the middle of the tool bar called **Two-Way Transceiver/Appliance Module, RR501**.

NOTE: To see exactly what module type is depicted by the icons in the toolbar, simply hover the mouse pointer over the top of the different buttons.

18. Verify its photograph ID by clicking on the picture to make it disappear.

19. In the top "Representative Icon" area of the module, select the wall lamp with the up-shade so that the module appears as shown in Figure 1-13.

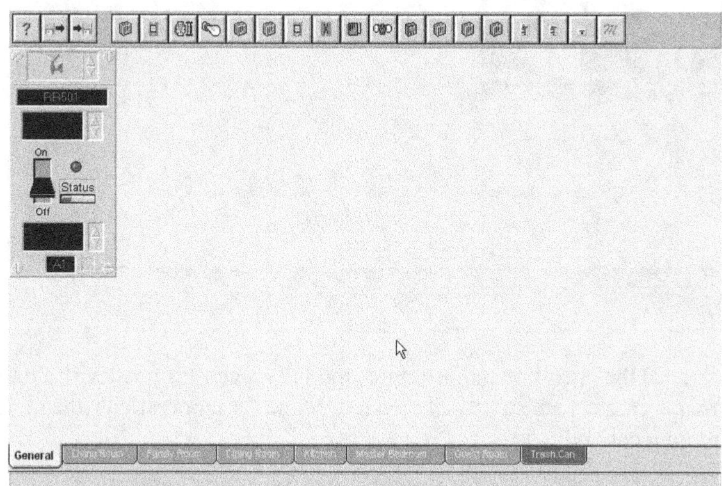

Figure 1-13: Two-Way Transceiver/Appliance Module on Screen

20. Click the icon on the left of the toolbar called **Lamp Module, LM465**.

21. Verify its photograph ID by clicking on the picture to make it disappear.

22. In the top "Representative Icon" area of the module, select another wall lamp with the up-shade so that the screen work area appears as shown in Figure 1-14.

23. Click on another **Lamp Module, LM465** from the toolbar.

24. Verify its photograph ID by clicking on the picture to make it disappear.

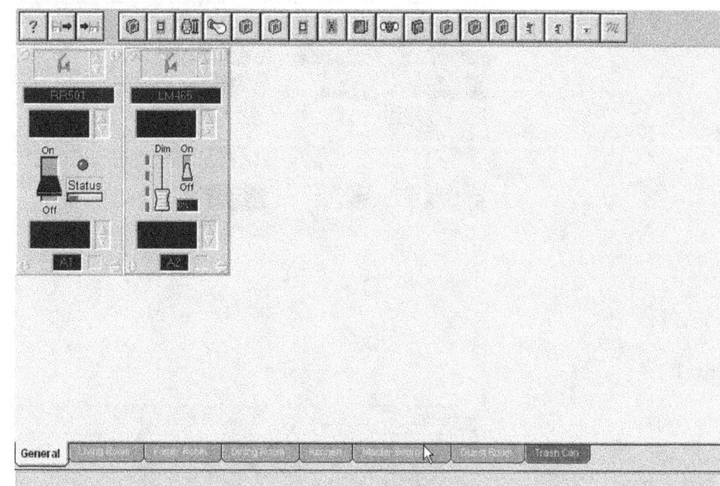

Figure 1-14: Lamp Module Using Up-Shade Lamp Icon

25. In the top "Representative Icon" area of the module, select a third wall lamp with the up-shade.

26. At the bottom of this module, type in the address *A4*, so that the screen appears as shown in Figure 1-15, and press **ENTER**.

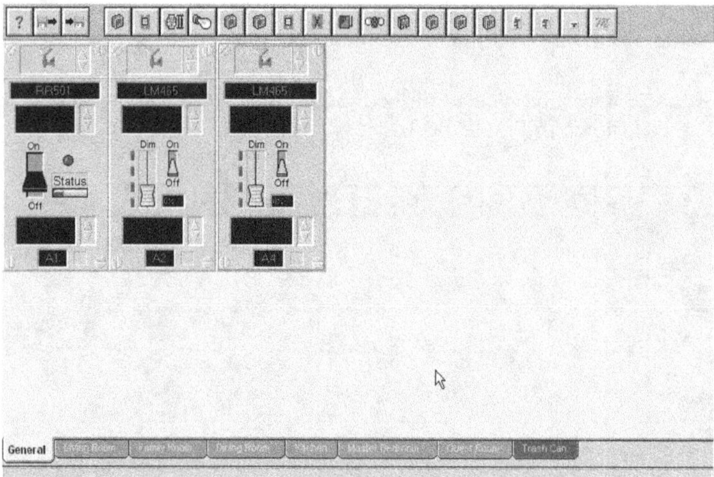

**Figure 1-15:
Transceiver and
Two Lamp
Modules on Screen**

At this point, you have programmed the operation mode of three modules, each assigned with a light to control. While the transceiver's control parameters are strictly limited to ON and OFF operations, the remaining modules each have additional dimming capabilities.

Testing Basic Power Control Program Operations

1. Move the cursor to the ON area of the switch on the "RR501" transceiver module icon assigned to address *A1*, as shown in Figure 1-16.

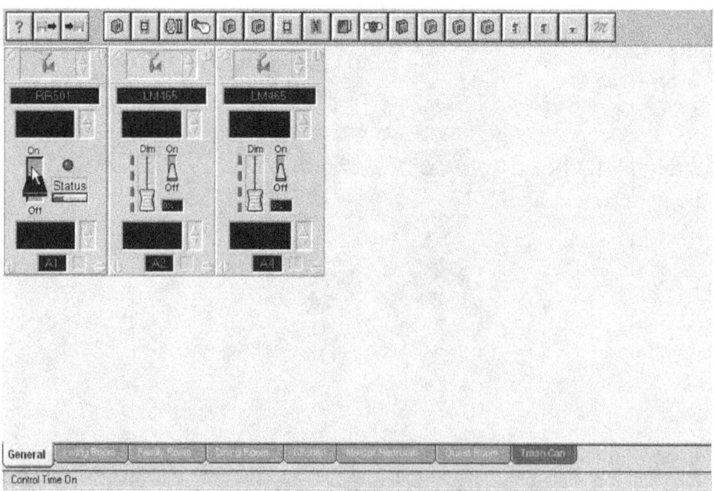

**Figure 1-16:
Transceiver Switch
ON Screen
Location**

2. Click once on the **ON** switch area in the "RR501" module icon.

The nightlight plugged into the transceiver module should come on. Although the transceiver has a selectable house code (currently set to *A*), its unit code is always *1*.

3. Click the **OFF** switch area of the "RR501" module icon to extinguish the transceiver's nightlight.

4. Click on the **ON** switch area in the "LM465" module icon showing address *A2*.

The nightlight plugged into the lamp module at the bottom left of the control box should illuminate.

5. Move the cursor/pointer to the **DIM** slide control on the "LM465" module icon and click on it.

6. Then, move the slider to its halfway position and release it, as shown in Figure 1-17.

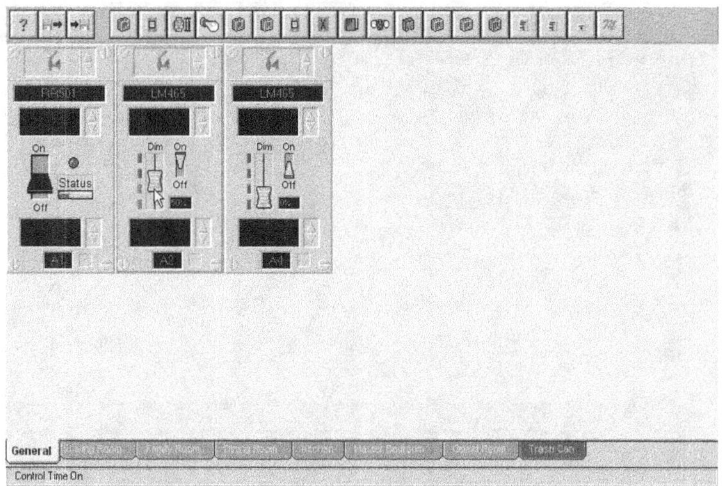

Figure 1-17:
Moving the DIM
Slider to its
Halfway Position
on the LM486 Icon

The nightlight plugged into the lamp module at the bottom left of the control box should dim to about half its full intensity.

7. Click once on the **OFF** switch area of the "LM465" module icon to extinguish the nightlight plugged into the lamp module at the bottom left of the control box.

8. Click on the **ON** switch area in the "LM465" module icon showing address *A4*.

The nightlight plugged into the lamp module at the top right of the control box should illuminate.

9. Move the cursor/pointer to the DIM slide control on the "LM465" module icon and try various settings to verify that the nightlight can be effectively controlled.

10. Click once on the **OFF** switch area of the "LM465" module icon to extinguish the nightlight plugged into the lamp module at the top right of the control box.

11. Verify that the key chain remote control and the 6-in-1 remote control also operate the modules you've created, and that the computer images react to all operations.

A slight delay will occur between the time a command is executed at a remote control and a visual confirmation is observed on the computer, so give the program enough time to visually react to each command signal. The actual modules on the Marcraft panel will react more quickly to operations directed to them by the remote controls.

12. Use any method of choice to turn all modules **OFF**.

13. Save the programming worksheet you've just created by first inserting a 3.5-inch disk into the floppy drive on the computer.

14. On the toolbar click on the **Save file to disk** command button, as shown in Figure 1-18.

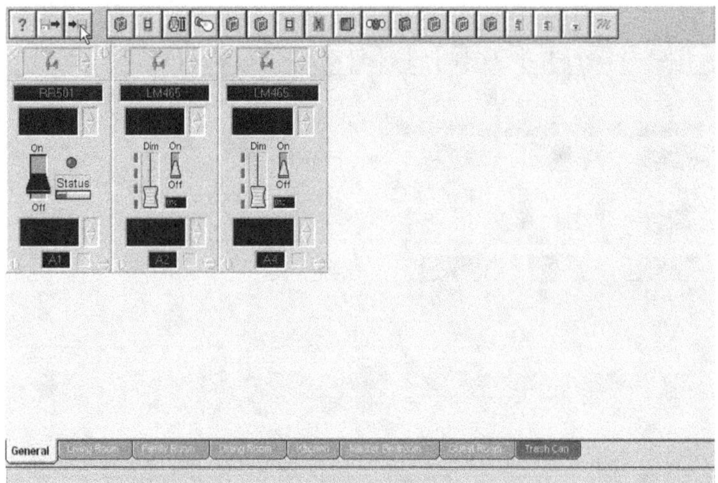

**Figure 1-18:
Executing the
Save File to Disk
Command**

15. When the *Save As* dialog box appears, click on the **Drives:** drop-down list and select the **a:** drive as the destination.

16. Enter **pgm1xxxx.X10** as the filename, where your initials and those of your lab partner are substituted for *xxxx*.

17. From the dialog box shown in Figure 1-19, click on the **OK** button to save the file to your floppy disk.

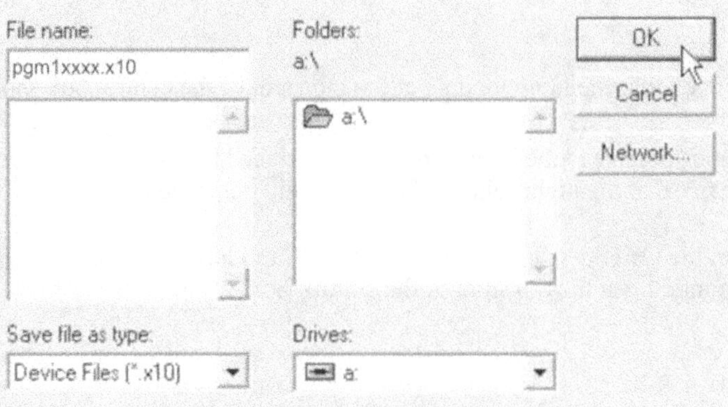

**Figure 1-19:
Saving the
Program
Worksheet**

18. Remove your work disk from the floppy drive, label it, and store it in a safe location.

19. Remove power to the Marcraft panel by moving its power toggle switch (lower right) to its **OFF** (right) position.

20. Close the ActiveHome software program.

21. Close Windows and turn the computer system **OFF**.

LAB QUESTIONS

Feedback

1. Why didn't the lamp plugged into the transceiver module activate when the *POWER* button on the 6-in-1 remote control was pressed?

2. What happens when a de-energized appliance module receives either a *VOLUME+* or a *VOLUME–* code?

3. What happens when an X-10 house code on a module does not match the house code sent to/from the transceiver?

4. How is the work area of the ActiveHome program screen cleared?

5. How did you extinguish the nightlight at module *A2* after manually moving its toggle switch first OFF, and then ON?

6. What happens when you press the pushbutton of the inactive wall switch module assigned to *A5*?

7. What happens to the 120 Vac fan unit when the 6-in-1 remote's *6* button is pressed, and then its *VOLUME–* rocker button?

8. What does the opening screen for the ActiveHome software program look like?

9. What does the ActiveHome software program call the toolbar's RR501 icon?

10. Which module was the key chain remote control unable to operate? Why?

A Multizone X-10 Power and Control System

OBJECTIVES

1. Design a multizone residential power control system.
2. Implement a multizone residential power control system.
3. Test and verify multizone residential power control system operations.

Power and Control Systems

RESOURCES

1. Marcraft Residential Power and Control Experiment Panel and Frame
2. X-10 Powerhouse ActiveHome Owner's Manual
3. Computer system with Windows XP Professional and X-10 Powerhouse ActiveHome software installed
4. Floppy disk, 3.5" (work disk)
5. Pencil
6. Paper

DISCUSSION

Most home automation installations are organized using a zoning system. Various zones are often created and activated using devices that perform similar operations, or modules that perform dissimilar actions at identical times. Zoning makes it easier to troubleshoot systems when they begin having problems. Remember that the division of modules into different zones does not in any way correspond to the circuit arrangement in the control box. As long as the phase coupler is doing its job, control signals should have no problem reaching their intended target devices.

In order to implement a zone power control system, as with other residential systems, the RESI Environmental Control technician should consult closely with the client about what operations are expected. Devices that are grouped together into a zone will not necessarily share identical address identifiers, although they might under certain circumstances. On the Marcraft panel, the individual modules will retain their unique address identifiers throughout this procedure.

Suppose that a zoning plan divides the available modules on the Marcraft panel into three zones. Zone 1 could be reserved for fan operations. Zone 2 might handle the normal after-dark lighting requirements. Zone 3 might light a room that remains completely shielded from natural daylight.

PROCEDURES

Designing a Multizone Power Control System

1. Use pencil and paper to create a two-column/eight row table.

2. Label the top row **ORGANIZATIONAL ZONING PLAN**.

3. In the second row, title the left column **ZONES** and the right column **OPERATIONS**.

4. Under the *ZONES* heading, combine the next three rows together and label them as **Zone 1**.

5. Below the Zone 1 rows under the *ZONES* heading, combine the next two rows together and label them as **Zone 2**.

6. Under the *ZONES* heading, label he bottom row **Zone 3**.

The *ZONES* column organizes three separate zones for the modules mounted on the Marcraft panel.

7. Under the *OPERATIONS* heading, label the third row **Transceiver Light**.

8. Under the *OPERATIONS* heading, label the fourth row **120VAC Fan Unit**.

9. Under the *OPERATIONS* heading, label the fifth row **240VAC Lamp**.

Zone 1 will consist of the 120 Vac fan and a 240 Vac lamp, along with the nightlight plugged into the transceiver. The light will signal when either of these fans has been activated. As such, the transceiver light indicates that air filtering is taking place. Air filtering would occur most likely during the early morning when bathroom use is fairly heavy, or in the evening after dinner has been served. The 240 Vac lamp will be located in the kitchen, while the 120 Vac fan will service the shower area.

10. Under the *OPERATIONS* heading, label the sixth row **Lamp Module 1**.

11. Under the *OPERATIONS* heading, label the seventh row **Lamp Module 2**.

Zone 2 will consist of two nightlights plugged into their respective lamp modules. They will provide the residential lighting required during normal hours of darkness.

12. Under the *OPERATIONS* heading, label the eighth row **Ceiling Light Switch**.

Zone 3 will consist solely of the ceiling light attached to the wall switch circuit. If this light were situated in a windowless recreation room, its operation would be required whenever the room was occupied, although certainly not always at its maximum setting.

13. Compare your organizational zoning plan with the one shown in Table 2-1.

Table 2-1: Organizational Zoning Plan

Zones	Operations
Zone 1	Transceiver Light
	120 Vac Fan Unit (shower)
	240 Vac Lamp Unit (kitchen)
Zone 2	Lamp Module 1 Dimmer
	Lamp Module 2 Dimmer
Zone 3	Ceiling Light Switch

14. After examining Table 2-1 carefully, make any required changes or adjustments to your organizational zoning plan.

Implementing a Multizone Residential Power Control System

1. Turn the computer system **ON** and wait for the *Windows XP* screen to appear.

2. Launch the ActiveHome software program and click on the **MAXIMIZE** button in the upper-right corner of the program window to fill the available screen area.

3. Clear the work area of the screen by pressing the **F2** function key, or selecting the *New* command from the File menu listing.

4. Make sure you have a copy of your organizational zoning plan within view.

5. From the toolbar, click on the icon labeled **Two-Way Transceiver/Appliance Module, RR501**.

6. Click on the picture of the transceiver module to close it.

7. In the *Unit Descriptor* window, label the module **Zn1 Transcvr Lt**.

8. From the toolbar, click on the icon labeled **Receptacle Module, SR227**.

9. Click on the picture of the receptacle module to close it.

10. In the *Unit Descriptor* window, label the module **Zn1 Shower Fan**.

11. In the *Module Housecode Unit Code* ID box, enter **A6**.

12. From the toolbar, click on the icon labeled **220V, 20A Heavy Duty Appliance Module, HD245**.

13. Click on the module's picture to close it.

It's OK if the picture does not exactly match the 240 Vac receptacle mounted on the Marcraft panel.

14. In the *Unit Descriptor* window, label the module **Zn1 Kitchen Lamp**.

15. In the *Module Housecode Unit Code* ID box, enter **A3**.

16. From the toolbar, click on the icon labeled **Show modules in minimum form**.

At this point, your worksheet should appear similar to that shown in Figure 2-1.

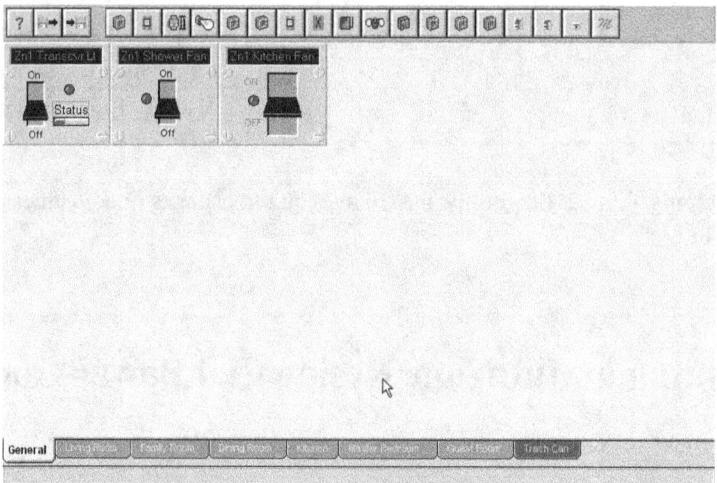

Figure 2-1:
Multizone Program
Worksheet for Zone 1

17. From the toolbar, click on the icon labeled **Lamp Module, LM 465**.

18. Click on the module's picture to close it, and in its *Module Housecode Unit Code* ID box, enter **A2** if necessary.

19. In the *Unit Descriptor* window, label the module **Zn2 Lamp 1**.

20. From the toolbar, click on the icon labeled **Show modules in minimum form**.

21. From the toolbar, click on the icon labeled **Lamp Module, LM 465** again.

22. Click on the module's picture to close it, and in its *Module Housecode Unit Code* ID box, enter **A4**.

23. In the *Unit Descriptor* window, label this lamp module **Zn2 Lamp 2**.

24. From the toolbar click on the icon labeled **Show modules in minimum form**.

25. From the menu, click on the **Modules/Lamp Module/Wall Switch Module WS467** selection.

26. Click on the module's picture to close it, and in its *Module Housecode Unit Code* ID box, enter **A5**, if necessary.

27. In the *Unit Descriptor* window, label this lamp module **Zn3 Ceiling Light**.

28. From the toolbar, click on the icon labeled **Show modules in minimum form**.

If the unit descriptor does not display its new label, click once in its window and once in an unoccupied region of the screen.

29. Use the mouse to individually grab each module for Zone 2 and place them below the first row of modules, at the far left.

30. Use the mouse to grab the module for Zone 3 and place it below the first two rows of modules, at the far left.

The screen should now appear similar to that shown in Figure 2-2.

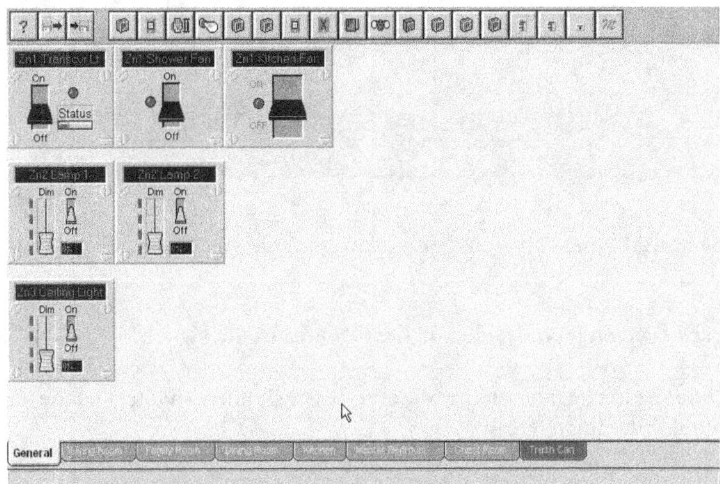

Figure 2-2: Multizone Module Placement

31. Save the programming worksheet you've just created by inserting your 3.5-inch disk into the floppy drive on the computer.

32. On the toolbar, click on the **Save file to disk** command button.

33. When the *Save As* dialog box appears, click on the **Drives:** drop-down list and select the **a:\ drive** as the destination.

34. Enter **pgm2xxxx.x10** as the filename, where your initials and those of your lab partner are substituted for *xxxx*.

35. From the dialog box, click on the **OK** button to save the file to your floppy disk.

You have successfully implemented a multizone residential power control system. You will now test the system for operability.

Testing and Verifying Multizone Residential Power Control System Operations

1. Make sure that *pgm2xxxx.x10* is still loaded in the ActiveHome software program, where your initials and those of your lab partner are substituted for *xxxx*.

2. If necessary, load it and check for a screen similar to that shown in Figure 2-3.

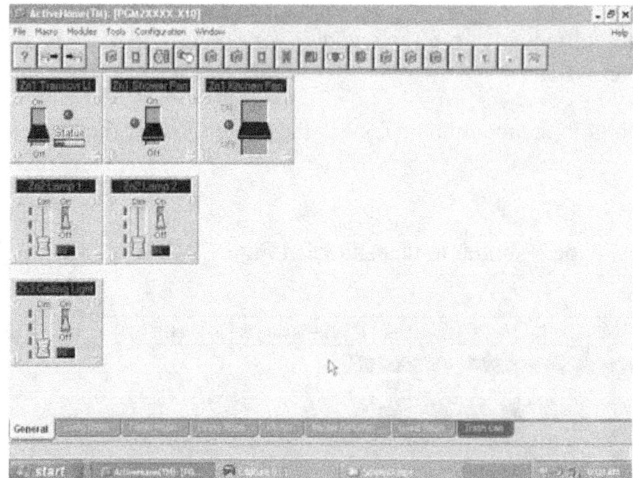

Figure 2-3:
Program File
pgm2xxxx.x10
Loaded

3. Apply power to the Marcraft panel by moving its power toggle switch (lower right) to its **ON** (left) position.

4. At the *Zn1 Transcvr Lt* screen module, click in the **ON** area of the switch.

The nightlight plugged into the transceiver on the Marcraft panel should activate, and the *Status* indicator for the on-screen module should illuminate green.

5. At the *Zn1 Shower Fan* screen module, click in the **ON** area of the switch.

Now the 120 Vac fan unit activates, while the transceiver nightlight remains lit. The *Status* indicators for both on-screen modules are illuminated green.

6. At the *Zn1 Kitchen Lamp* screen module, click in the **ON** area of the switch.

Now the 240 Vac lamp unit activates, while the 120 Vac fan continues to run and the transceiver nightlight remains lit. The *Status* indicators for all three Zone 1 on-screen modules are illuminated green.

7. At the 6-in-1 remote control unit, press the **MUTE** button.

The fan unit, the kitchen lamp, and the transceiver nightlight are turned OFF, with corresponding visual confirmations at the applicable on-screen *Status* indicators.

8. Click in the **ON** switch control area of the *Zn2 Lamp 1* module.

The nightlight plugged into the lamp module at the lower left of the Marcraft panel fully illuminates, while the on-screen LED indicators verify the actual intensity level of the bulb.

9. Click in the **ON** switch control area of the *Zn2 Lamp 2* module.

The nightlight plugged into the lamp module at the upper right of the Marcraft panel also illuminates. On-screen LED indicators verify full intensity levels for both lamp module bulbs.

10. At the on-screen *Dim* controls for these lamps, verify the individual operation of their sliders can be established, and that their LED indicators accurately report their panel status.

11. Also at the 6-in-1 remote control unit, verify that each lamp can be manipulated through various degrees of intensity. Then, press the **MUTE** button.

Both lamp units are turned OFF, with corresponding visual confirmations at the applicable on-screen *Status* indicators.

> ## CAUTION
>
> Avoid looking directly at the ceiling light's 40-watt lamp! Even though it is only 40 watts, it is still strong enough to cause a spot in your vision for several minutes if you stare at it.

12. Click in the **ON** switch control area of the on-screen *Zn3 Ceiling Light* module.

The ceiling light illuminates fully ON, with corresponding visual confirmations at the applicable on-screen LED indicators.

13. At the on-screen *Dim* control for this light, verify the operation of its slider, and that its on-screen LED indicators accurately report the panel status.

14. Also at the 6-in-1 remote control unit, verify that the ceiling light can be turned **ON** and **OFF**, with matching indications at the on-screen module indicators.

15. With the ceiling light turned *ON* at full illumination, use the **VOLUME–** rocker on the 6-in-1 remote to move the light's illumination level to near **50%**.

16. Use the **VOLUME+** rocker on the 6-in-1 remote control to increase the dimmer's setting to maximum intensity for the ceiling light.

17. Then, use the remote's **CHANNEL–** rocker to turn the ceiling light **OFF**.

18. Now, use any method to turn the ceiling light back **ON**.

19. Finally, click in the **OFF** switch control area of the on-screen *Zn3 Ceiling Light* module.

20. Remove your work disk from the floppy drive, and store it in a safe location.

21. Remove power to the Marcraft panel by moving its power toggle switch (lower right) to its **OFF** (right) position.

22. Close the ActiveHome software program.

23. Close Windows and turn the computer system **OFF**.

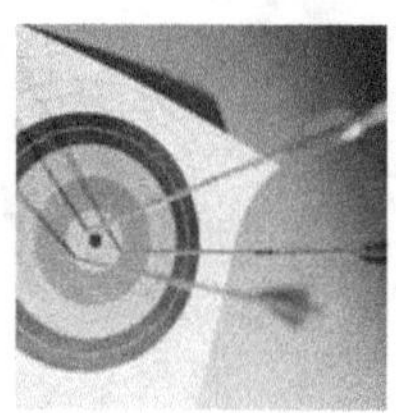

LAB QUESTIONS

1. How many zones were created for the organizational zoning plan?

2. What activities were primarily reserved for Zone 1 in the organizational zoning plan?

3. How was the SR227 receptacle module labeled in its Unit Descriptor window?

4. What household code is used by the kitchen lamp in Zone 1?

5. Why does the wall switch module controlling the ceiling light refuse to operate through the normal use of its pushbutton?

6. What do the on-screen LED indicators for the nightlight lamp modules indicate?

7. What is the result of using the 6-in-1 remote control's CHANNEL– rocker?

8. How many different modules are controlled by the *pgm2xxxx.x10* program file?

An Automated Multizone X-10 Power and Control System

OBJECTIVES

1. Design multizone automated residential power control system features.
2. Implement the operation of automated multizone residential power control devices.
3. Test and verify automated multizone power control device operations.

Power and Control Systems

RESOURCES

1. Marcraft Residential Power and Control Experiment Panel and Frame
2. X-10 Powerhouse ActiveHome Owner's Manual
3. Computer system with Windows XP Professional and X-10 Powerhouse ActiveHome software installed
4. X-10 Powerhouse ActiveHome Automation System CK11A
5. Floppy disk, 3.5" (work disk)

DISCUSSION

Because zoned power and control systems adapt to specific conditions existing at a particular residence, careful planning always precedes the partitioning of the various zones. Recall from the previous lab procedure the zoning plan that resulted in the creation of the *pgm2xxxx.x10* program file, which was saved to your work disk.

In that program file, Zone 1 consisted of the 120 Vac fan unit and a 240 Vac lamp unit, along with the transceiver nightlight. The initial idea was to use the light to signal when either fan was running, or at least when ventilation was supposed to take place. The discussion suggested that automated air filtering would most likely occur during early morning and late evening time periods. The plan envisioned using the 120 Vac fan unit in the shower area, and the 240 Vac lamp unit as a kitchen light.

Zone 2 used two lamp modules to provide nighttime residential lighting. Zone 3 consisted of a ceiling light attached to the wall switch circuit, such as might be found in a recreation room that does not have natural light. If this light were situated in a windowless environment, it would usually be operated whenever the room was occupied. Light levels would vary depending on the programming from system software commands.

**Power and
Control Systems**

PROCEDURES

Designing Multizone Automated Residential Power Control System Features

Previous lab activities have verified the active link between the Marcraft panel and the ActiveHome computer program, but an automated program has not yet been designed. You will now design a multizone power control system with automated features. A normal programming routine would take a full 24-hour time sequence into account, but the classroom environment requires that a sequence occur over a period of minutes, rather than hours.

1. Turn the computer system **ON** and wait for the *Windows XP* screen to appear.

2. Launch the ActiveHome software program and click on the **MAXIMIZE** button in the upper-right corner of the program window to fill the available screen area.

3. Clear the work area of the screen by pressing the **F2** function key, or selecting the *New* command from the File menu.

4. Insert your 3.5" work disk into the floppy drive.

5. Locate the **pgm2xxxx.x10** program file you created earlier and load it into the ActiveHome software program.

6. At the menu bar click on the **File/Save as...** command.

7. In the *Save As* dialog box, select the **a:** drive and type **pgm3xxxx.x10** for the *File name:* entry, as shown in Figure 3-1.

**Figure 3-1:
Saving a
pgm3xxxx.x10
Work File**

8. Click on the **OK** button to save the file as *pgm3xxxx.x10*.

9. Observe the sheet tabs at the bottom of the *ActiveHome* program screen. You should currently be located in the **General** layout sheet.

10. Double-click on the **Family Room** sheet tab, if it exists.

11. When the title text becomes highlighted, type **Recreation Room**.

12. Now, double-click on the **Guest Room** sheet tab, if it exists.

13. When the title text becomes highlighted, type **Bathroom**.

14. Return to the program screen's **General** layout sheet for *pgm3xxxx.x10*, which should now appear as shown in Figure 3-2.

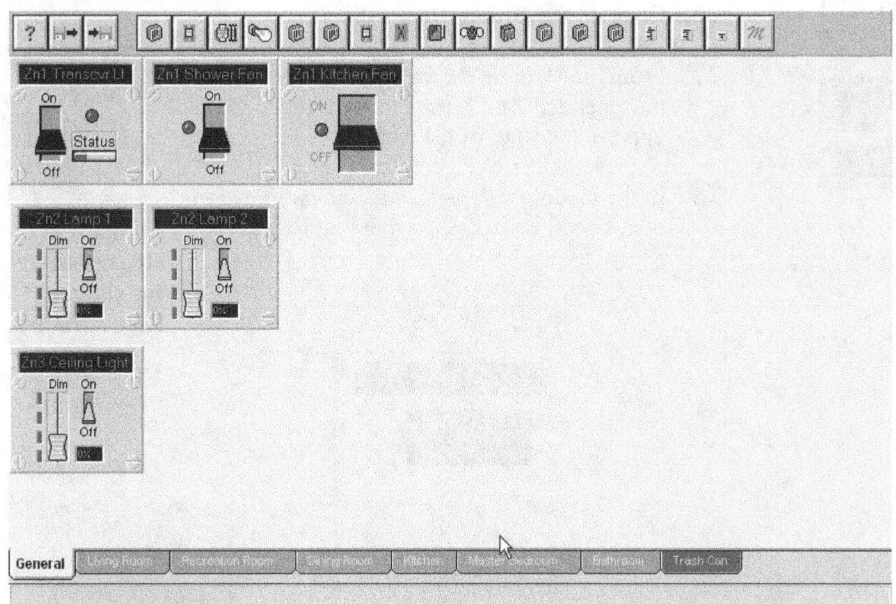

**Figure 3-2:
Layout Sheet with
Renamed Tabs**

15. Click and drag the **Zn3 Ceiling Light** module icon onto the area just above the "Recreation Room" sheet tab, and click on the **Yes** button in the *Confirmation* box to relocate it.

16. Click and drag the **Zn2 Lamp 1** module icon onto the area just above the "Dining Room" sheet tab, and click on the **Yes** button in the *Confirmation* box to relocate it.

17. Click and drag the **Zn2 Lamp 2** module icon onto the area just above the "Master Bedroom" sheet tab, and click on the **Yes** button in the *Confirmation* box to relocate it.

18. Click and drag the **Zn1 Kitchen Lamp**, the **Zn1 Shower Fan**, and the **Zn1 Transcvr Lt** module icons onto the areas just above the "Kitchen", "Bathroom", and "Living Room" sheet tabs respectively. Relocate each module icon using its respective *Confirmation* box.

You can also move module icons to the various sheet tabs by dragging each over its target sheet tab and then clicking once on the target. When you have finished, the "General" sheet tab will be empty.

19. Click on the **Living Room** sheet tab.

20. From the "Living Room" sheet layout click on the **Show modules in extended form** tool button.

21. Move the extended **Zn1 Transcvr Lt** module icon to the upper-left corner of the worksheet.

22. In the module's *Representative Icon* selection box, select the icon shown in Figure 3-3.

23. Click on the **Recreation Room** sheet tab.

24. From the "Recreation Room" sheet layout, move the extended **Zn3 Ceiling Light** module icon to the upper-left corner of the worksheet.

25. In the module's *Representative Icon* selection box, select the icon shown in Figure 3-4.

26. Click on the **Dining Room** sheet tab.

27. From the "Dining Room" sheet layout, move the extended **Zn2 Lamp 1** module icon to the upper-left corner of the worksheet.

28. In the module's *Representative Icon* selection box, select the icon shown in Figure 3-5.

Figure 3-3: Representative Icon for the Zn1 Transcvr Lt Module

Figure 3-4: Representative Icon for the Zn3 Ceiling Light Module

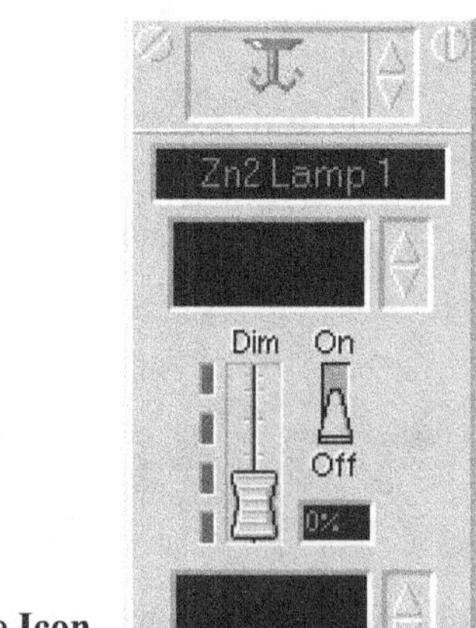

Figure 3-5: Representative Icon for the Zn2 Lamp 1 Module

29. Click on the **Kitchen** sheet tab.

30. From the "Kitchen" sheet layout, move the extended **Zn1 Kitchen Lamp** module icon to the upper-left corner of the worksheet.

31. In the module's *Representative Icon* selection box, select the icon shown in Figure 3-6.

32. Click on the **Master Bedroom** sheet tab.

33. From the "Master Bedroom" sheet layout, move the extended **Zn2 Lamp 2** module icon to the upper-left corner of the worksheet.

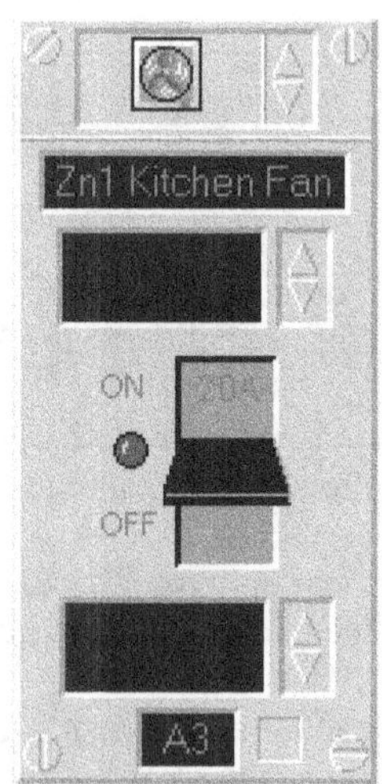

Figure 3-6: Representative Icon for the Zn1 Kitchen Lamp Module

34. In the module's *Representative Icon* selection box, select the icon shown in Figure 3-7.

35. Click on the **Bathroom** sheet tab.

36. From the "Bathroom" sheet layout, move the extended **Zn1 Shower Fan** module icon to the upper-left corner of the worksheet.

37. In the module's *Representative Icon* selection box, select the icon shown in Figure 3-8.

All module icons are now located in the layout sheets representing their design locations within the residence. For the time being, the "General" layout sheet is empty.

38. Return to the **General** layout sheet by clicking on its tab.

39. From the Menu bar, click on the **Macro/New Standard Macro** item.

The *Macro Generator* screen appears, with all of the modules you've installed listed at the right side. These modules will be used to design the automated multizone power control system.

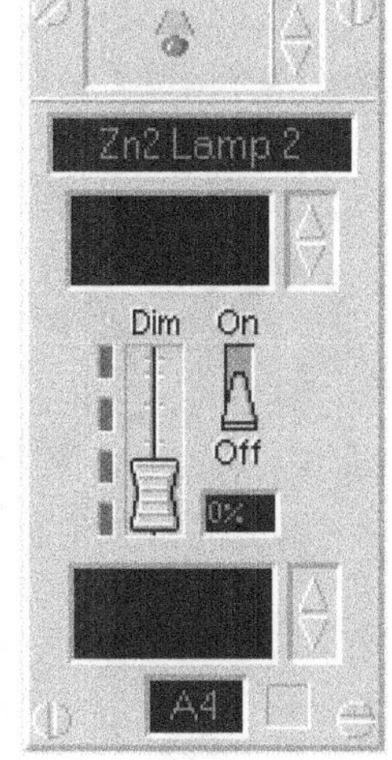

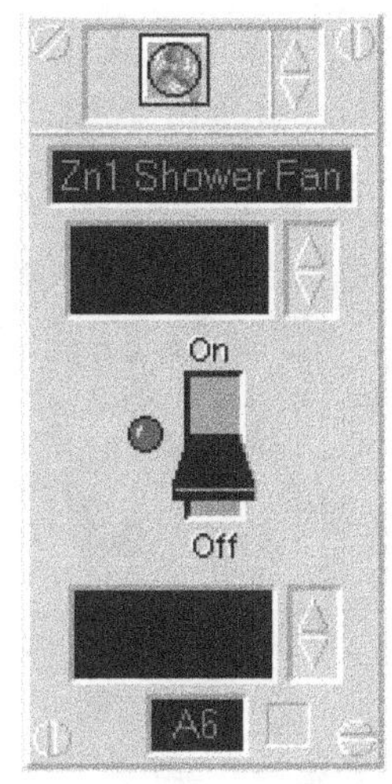

Figure 3-7: Representative Icon for the Zn2 Lamp 2 Module

Figure 3-8: Representative Icon for the Zn1 Shower Fan Module

40. At the top left of the screen, click in the **Macro initiator** window of the *Standard Macro placeholder* and type **B1**.

41. In the *Standard Macro placeholder*, click in the **Macro name** text box and type **Macroxxxx**, where xxxx are your initials and those of your lab partner.

The screen should look similar to that shown in Figure 3-9.

It's time to design a multizone power control system using automated features. Take into account the various activities that might normally take place in a residence over an entire 24-hour time sequence. In order to do this in the classroom, you will need to fit the entire program into a time period of approximately 30 minutes.

42. From the column of installed modules on the right of the screen, move the mouse pointer over the **Zn2 Lamp 2** to highlight its border with a red dotted line.

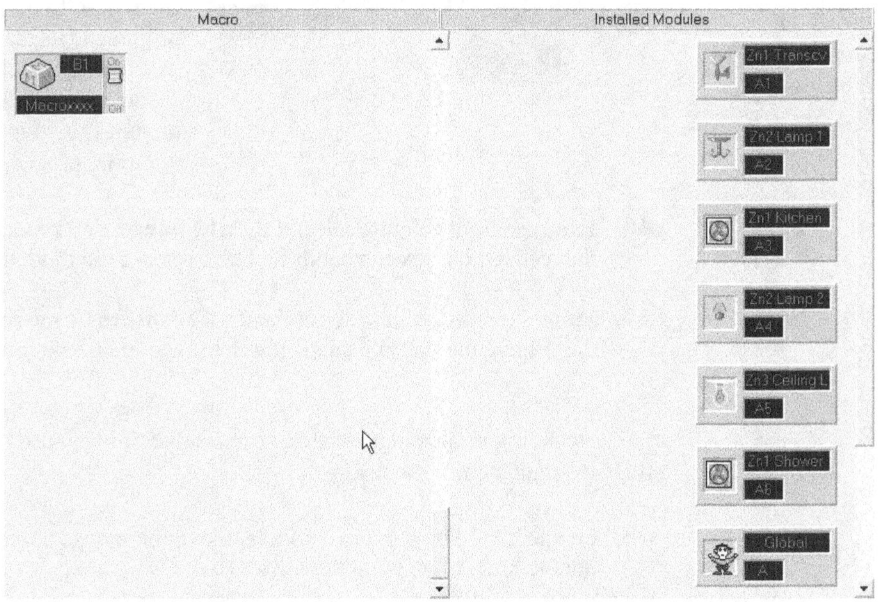

Figure 3-9: The Macro Generator Screen

43. Now click on the **Zn2 Lamp 2** module and drag it to the left of the screen until a rectangular box surrounds both it and the Macro placeholder.

44. Release the mouse button and drop the **Zn2 Lamp 2** module icon. It will lock into position beside the Macro placeholder, as shown in Figure 3-10.

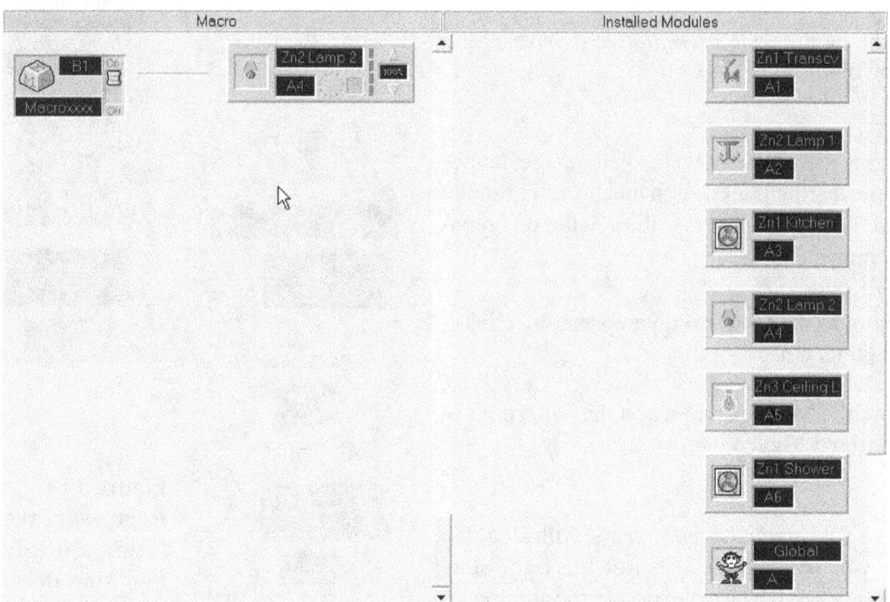

**Figure 3-10:
Placing a
Module Icon**

45. On the "Zn2 Lamp 2" module, the dimmer is already set to 100% brightness, as shown in Figure 3-11, with all four LED indicators lit.

**Figure 3-11:
Dimmer at 100%
Brightness**

To the left of the LED brightness indicators is a bright yellow square. This square determines the module's dimming mode. When the square is small, the module will simply dim from its current setting. When the square is large, the module will brighten to its maximum value before dimming. In order to easily observe which module the program is manipulating and to ensure accurate dimming instructions, we will instruct all light modules to go to maximum brightness prior to performing the dimming procedures.

46. From the right column, select the **Zn1 Shower Fan** module and drag it to the left of the screen until the rectangular box surrounds it. Then, release the module onto the work sheet.

47. From the right column, select the **Zn1 Transcvr Lt** module, and drag it to the left of the screen. When the rectangular box surrounds it, release the mouse button.

Because the nightlight plugged into the transceiver is being used as a visual signal for ventilation, it must be activated simultaneously with any fan.

48. On the "Zn1 Shower Fan" module, move the mouse pointer to the time delay function area and double-click to bring up the *Delay Time* dialog box.

49. In the *Delay Time* dialog box, drag the **green timer arrow** to the **one-minute** setting, similar to that shown in Figure 3-12.

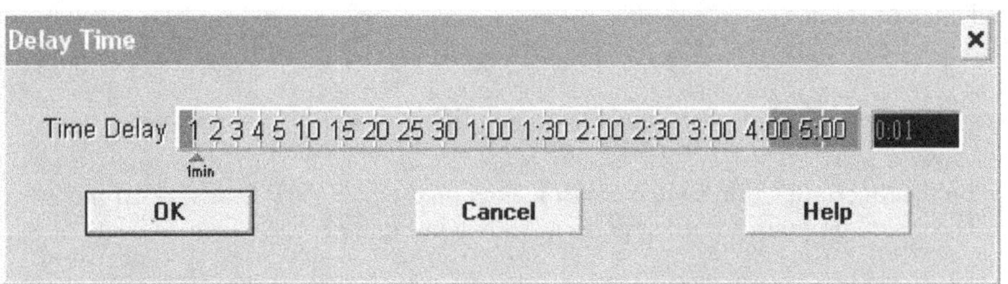

Figure 3-12:
Setting the Shower
Fan Delay Time

50. Then, click on the **OK** button to close the *Delay Time* dialog box.

Notice that the time delay setting now appears inside the clock area of the "Zn1 Shower Fan" module.

51. Set a time delay in the **Zn1 Transcvr Lt** module for exactly the same amount as you did for the "Zn1 Shower Fan" module.

Again, the nightlight plugged into the transceiver must mimic the ON and OFF actions of the ventilation fans. In this case, the nightlight will activate simultaneously with the shower fan. At this point, the Macro screen should appear as shown in Figure 3-13.

52. Drag another **Zn1 Shower Fan** module and another **Zn1 Transcvr Lt** module from the right column over to the left column, and add them to the event list.

If you try to release a module icon outside the rectangular inclusion box, the program assumes that you are starting a new macro, and will display the dialog box shown in Figure 3-14. To prevent this, always position each additional module within the displayed rectangular border before releasing it.

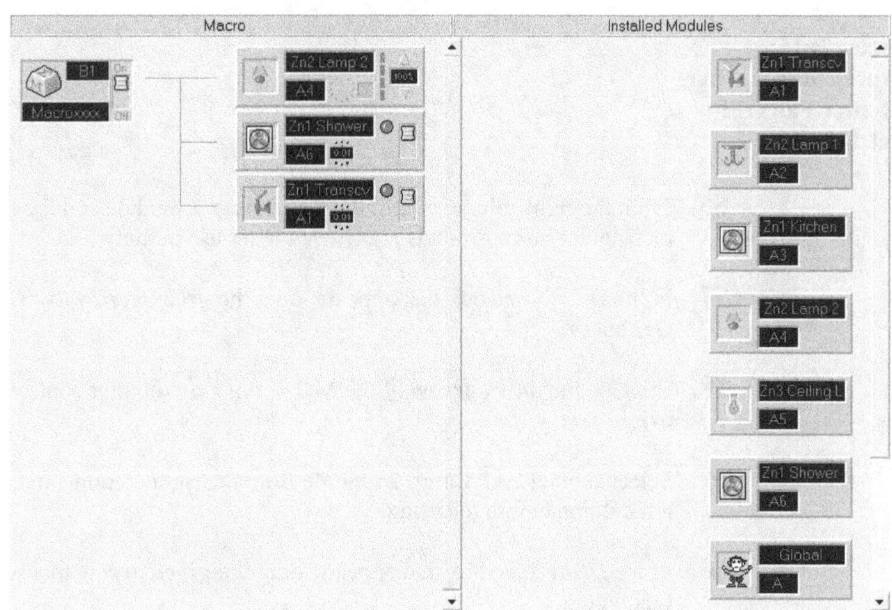

Figure 3-13:
Macro Screen After
Shower Fan Delay
Setting

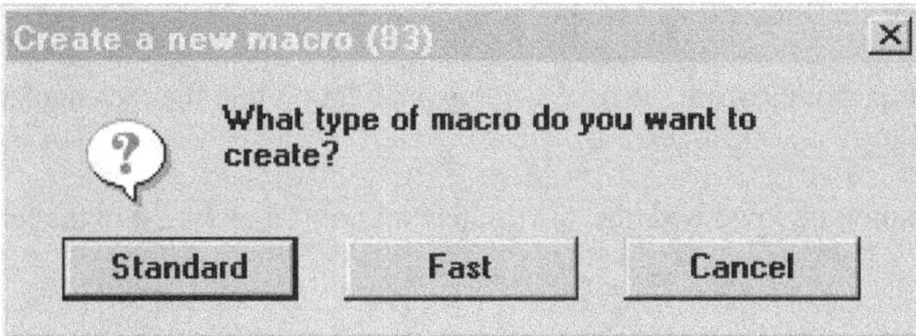

Figure 3-14:
New Macro
Dialog Box

53. In the new "Zn1 Shower Fan" and "Zn1 Transcvr Lt" modules, set the delay times for **two minutes**.

54. Now move the pointer to the Macro function switch control slider on the new **Zn1 Shower Fan** module and click on it once.

The slider moves down to the OFF position and the green LED indicator is extinguished.

55. Click once on the Macro function switch control slider on the new **Zn1 Transcvr Lt** module to turn it **OFF** also. The macro program screen appears as shown in Figure 3-15.

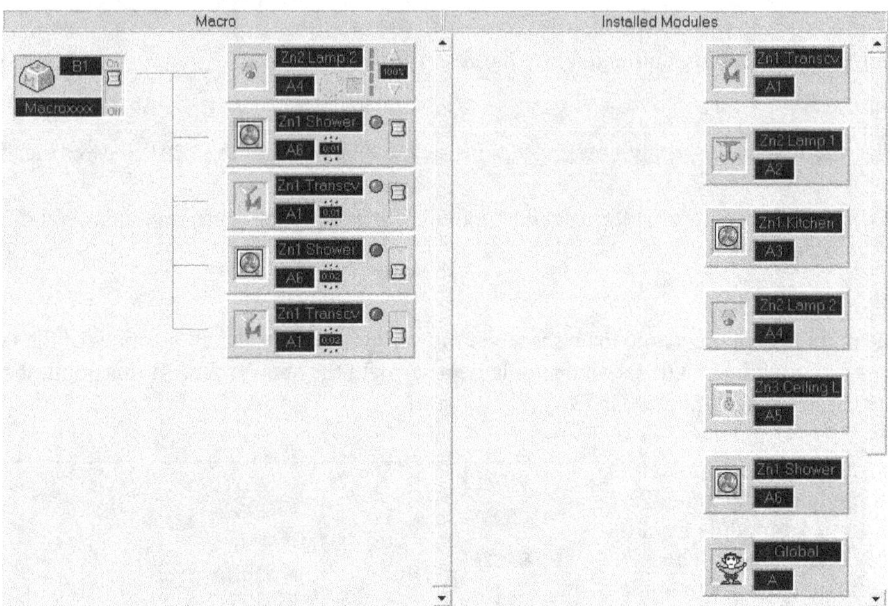

Figure 3-15:
Macro Screen After
Shower Fan OFF
Settings

56. From the right column, select the **Zn2 Lamp 2** module and drag it to the left of the screen. When the rectangular box surrounds it, release the mouse button.

57. In the *Delay Time* box that appears, drag the **green arrow** to a setting of **three minutes** and click the **OK** button.

58. Click on the **down arrow** of the Macro function dimmer control and set the dimmer for **60% brightness**.

59. Select another **Zn2 Lamp 2** module from the right column, and drag it into the rectangular box in the left column before releasing.

60. In the *Delay Time* box that appears, drag the **green arrow** to a setting, of **four minutes** and click the **OK** button.

61. Then, click repeatedly on the module's Macro function dimmer control's down slider to turn it completely **OFF**, with all four LED indicators extinguished.

62. Select the **Zn1 Kitchen Lamp** module from the right column, and drag it into the rectangular box in the left column before releasing.

63. Select the **Zn1 Transcvr Lt** module to go with the "Zn1 Kitchen Lamp" and drag it to the left column before releasing.

64. Set the delay times for the "Zn1 Kitchen Lamp" and "Zn1 Transcvr Lt" modules for **five minutes**.

Now the screen should look similar to that shown in Figure 3-16.

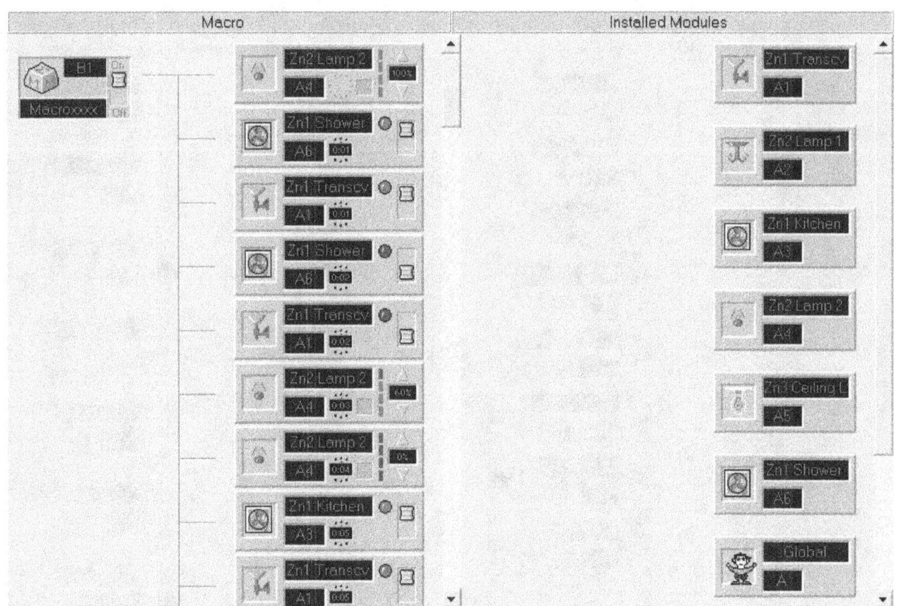

**Figure 3-16:
Macro Screen After
Kitchen Lamp and
Transceiver Module
Placement**

65. At this point, you need to make room on the screen for additional modules. Use the **down arrow** on the vertical scroll bar. Shift the macro screen up until the two most recently placed modules show at the top, as shown in Figure 3-17.

66. Drag a **Zn2 Lamp 1** module into the Macro list and set its delay time to **six minutes** and its brightness to **60%**.

To set the delay time to six minutes, you must use the text entry box at the far right of the time slider. Dragging the green arrow at this point would force a ten-minute delay instead.

67. Drag another set of **Zn1 Kitchen Lamp** and **Zn1 Transcvr Lt** modules into the Macro list.

68. Set their delay times to **seven minutes** and their function switches to **OFF**.

69. Drag a **Zn2 Lamp 1** module into the Macro list and set its delay time to **eight minutes**. Then, set its brightness to **0%**.

70. Drag a **Zn3 Ceiling Light** module to the list, set its delay time to **nine minutes**, and keep its brightness at **100%**.

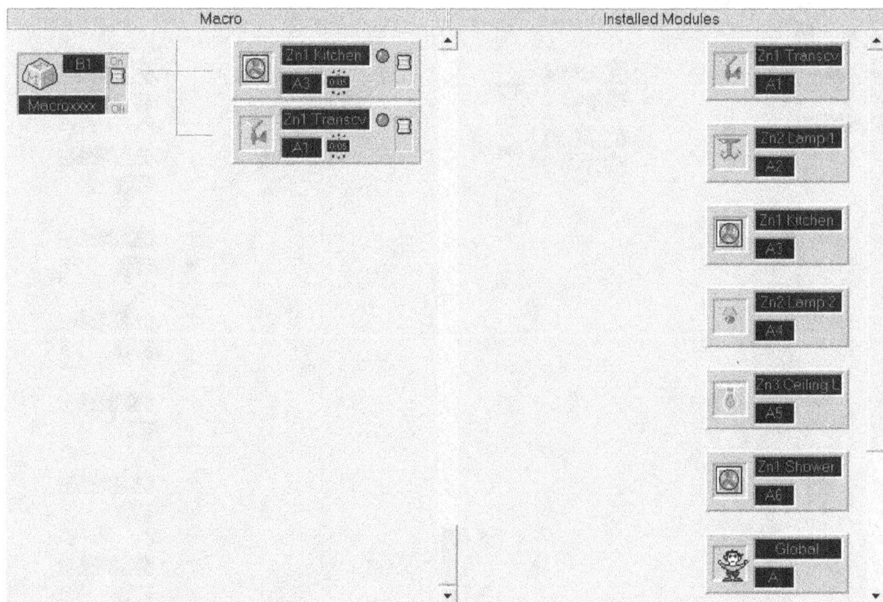

**Figure 3-17:
Macro Screen
After Scrolling
Up**

71. Drag another **Zn3 Ceiling Light** module to the list, set its delay time to **ten minutes**, and move its brightness to **0%**.

The screen should look similar to that shown in Figure 3-18.

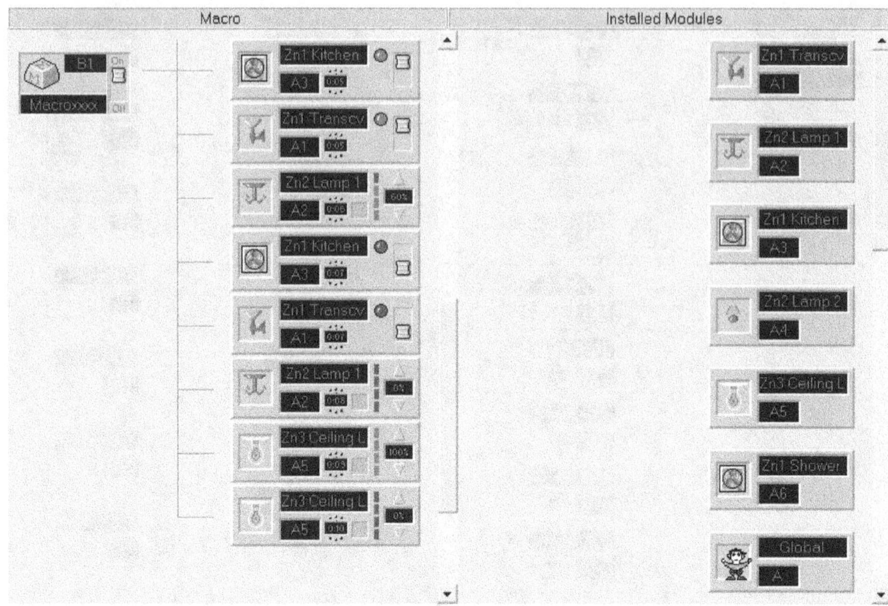

Figure 3-18:
Macro Screen After
Ceiling Light
Programmed OFF

72. Make room on the screen again for additional modules by using the **down arrow** on the vertical scroll bar. Shift the macro screen up until the two most recently placed modules show at the top, as shown in Figure 3-19.

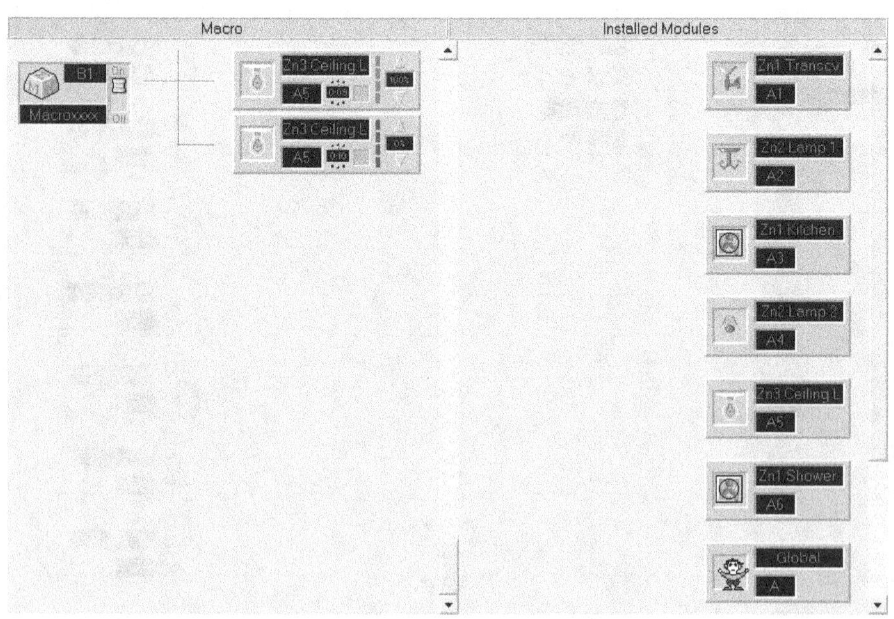

Figure 3-19: Macro Screen After
Scrolling Up the Second Time

73. Drag a set of **Zn1 Shower Fan** and **Zn1 Transcvr Lt** modules into position on the Macro list, and set both of their delay times for **eleven minutes**.

74. Drag another set of **Zn1 Shower Fan** and **Zn1 Transcvr Lt** modules into position on the Macro list, set both of their delay times for **twelve minutes**, and turn their switches **OFF**.

75. Drag a **Zn3 Ceiling Light** module into the list, set a delay time of **thirteen minutes**, and reduce its brightness to **60%**.

76. Drag another **Zn3 Ceiling Light** module into the list, set a delay time of **fourteen minutes**, and reduce its brightness to **0%**.

The screen should now look similar to that shown in Figure 3-20.

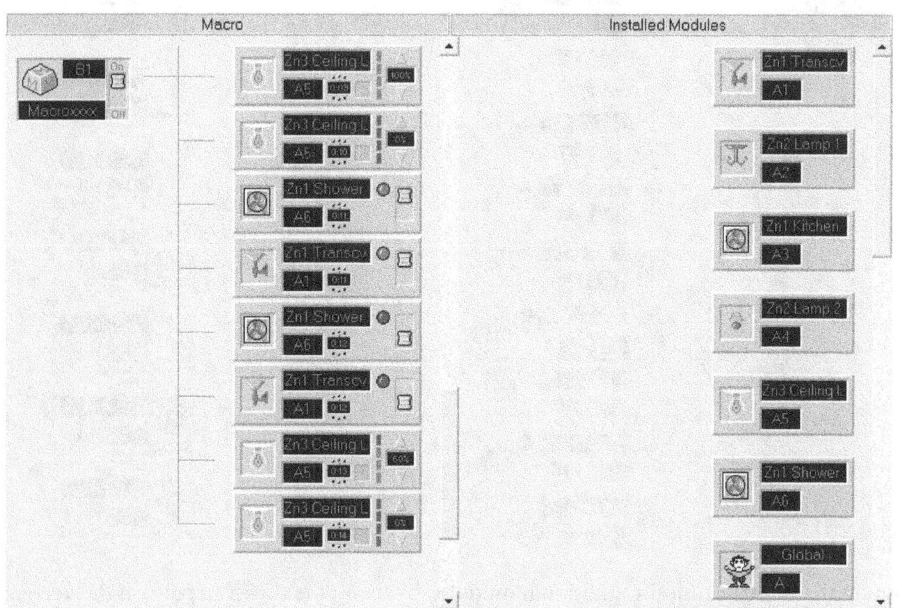

**Figure 3-20:
Macro Screen After
Ceiling Light
Programmed OFF**

77. Make room on the screen again for additional modules by using the **down arrow** on the vertical scroll bar. Shift the macro screen up until the two most recently placed modules show at the top, as shown in Figure 3-21.

78. Drag a set of **Zn1 Kitchen Lamp** and **Zn1 Transcvr Lt** modules into the list, and set a delay time of **fifteen minutes** for each. Function switches remain **ON** for both modules.

79. Drag a **Zn1 Shower Fan** module with an active function switch into the list, and assign it a delay time of **sixteen minutes**.

80. Drag another **Zn1 Shower Fan** module into the list, and assign it a delay time of seventeen minutes. Then, place its function switch in the **OFF** position.

**Figure 3-21:
Macro Screen After
Scrolling Up the
Third Time**

81. Drag a **Zn2 Lamp 1** module into the list, set its brightness to **100%**, and assign it a delay time of **seventeen minutes**.

82. Drag a set of **Zn1 Kitchen Lamp** and **Zn1 Transcvr Lt** modules into the list, and set a delay time of **eighteen minutes** for each. Function switches should be **OFF** for both modules.

The screen should now look similar to that shown in Figure 3-22.

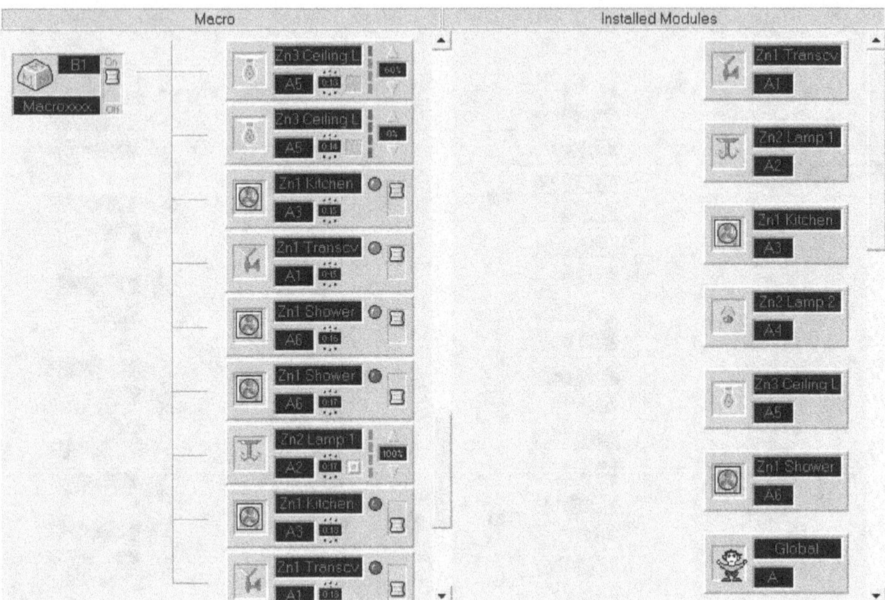

**Figure 3-22:
Macro Screen After
Kitchen Lamp
Programmed OFF**

83. Make room on the screen again for additional modules by using the **down arrow** on the vertical scroll bar. Shift the macro screen up until the two most recently placed modules show at the top, as shown in Figure 3-23.

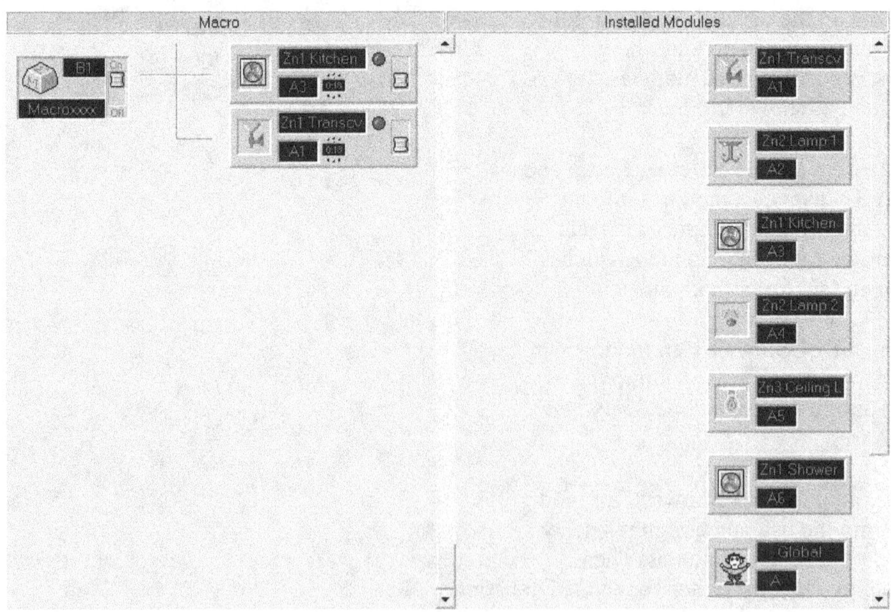

**Figure 3-23:
Macro Screen After
Scrolling Up the
Fourth Time**

84. Drag a **Zn2 Lamp 1** module into the list, set its brightness to **70%**, and assign it a delay time of **nineteen minutes**.

85. Drag a set of **Zn1 Kitchen Lamp** and **Zn1 Transcvr Lt** modules into the list, and set a delay time of **twenty minutes** for each. Function switches remain **ON** for both modules.

86. Drag a **Zn2 Lamp 1** module into the list, set its brightness to **60%**, and assign it a delay time of **twenty minutes**.

87. Drag a **Zn3 Ceiling Light** module into the list, and set a delay time of **twenty-one minutes** and a brightness of **100%**.

88. Drag a **Zn1 Shower Fan** module into the list, and set a delay time of **twenty-two minutes**.

89. Drag a **Zn1 Kitchen Lamp** module into the list, click the function switch to **OFF**, and set a delay time of **twenty-three minutes**.

The screen should now look similar to that shown in Figure 3-24.

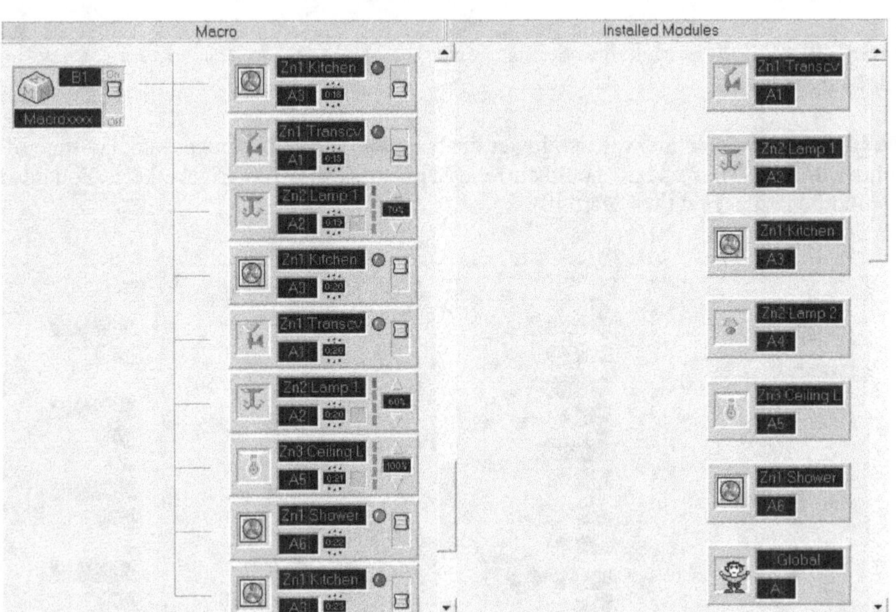

Figure 3-24: Macro Screen After Kitchen Lamp Programmed OFF at Twenty-Three Minutes

90. Make room on the screen again for additional modules by using the **down arrow** on the vertical scroll bar. Shift the macro screen up until the two most recently placed modules show at the top, as shown in Figure 3-25.

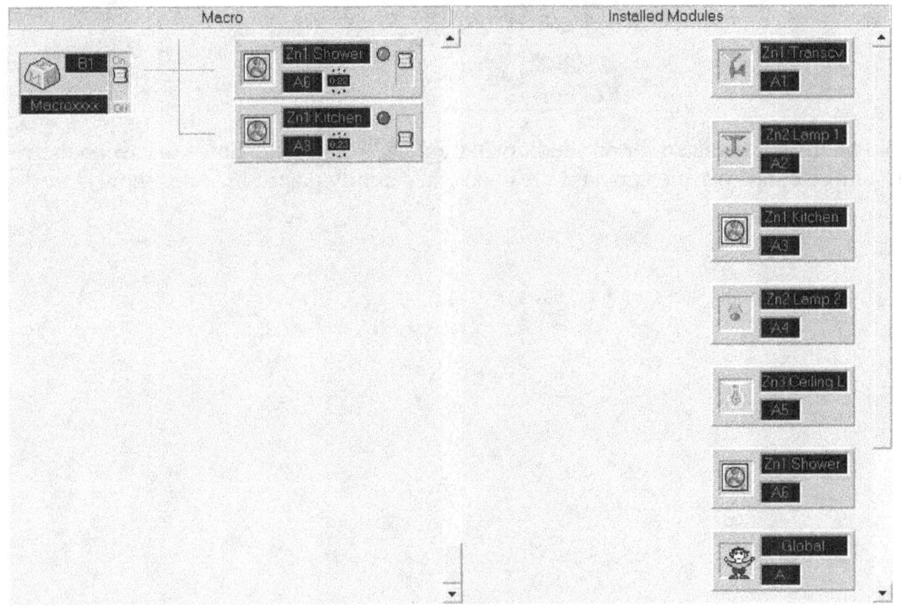

Figure 3-25: Macro Screen After Scrolling Up the Fifth Time

91. Drag a **Zn2 Lamp 1** module into the list, set the brightness to **0%**, and set a delay time of **twenty-four minutes**.

92. Drag a set of **Zn1 Shower Fan** and **Zn1 Transcvr Lt** modules into the list, and set a delay time of **twenty-five minutes** for each. Function switches should be switched OFF for both modules.

93. Drag a **Zn3 Ceiling Light** module into the list, set the brightness to **60%**, and set a delay time of **twenty-six minutes**.

94. Drag a **Zn2 Lamp 2** module into the list, keep the brightness at **100%**, and set a delay time of **twenty-seven minutes**.

95. Drag a **Zn3 Ceiling Light** module into the list, lower the brightness to **0%**, and set a delay time of **twenty-eight minutes**.

96. Drag a **Zn2 Lamp 2** module into the list, lower the brightness to **60%**, and set a delay time of **twenty-nine minutes**. Your screen should currently appear as shown in Figure 3-26. You now have one more module to place on the Macro list.

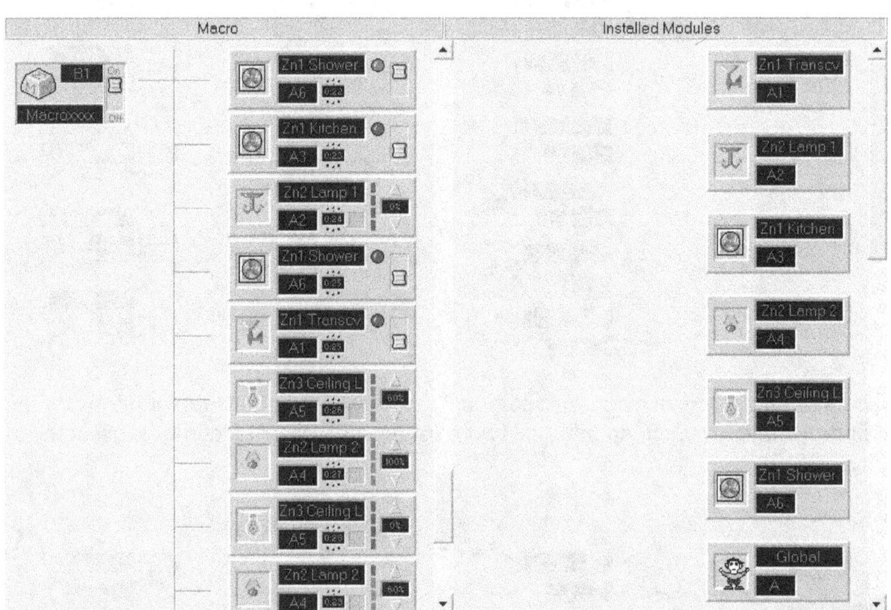

**Figure 3-26:
Macro Screen After
Lowering Lamp 2
Brightness**

97. Make room on the screen again for one additional module by using the **down arrow** on the vertical scroll bar. Shift the macro screen up until the two most recently placed modules show at the top.

98. Drag a final **Zn2 Lamp 2** module into the list, lower the brightness to **0%**, and set a delay time of **thirty minutes**.

You have just designed a multizone power control system using automated features. Your screen should currently appear as shown in Figure 3-27. When written as directed above, the entire program will take approximately 30 minutes to execute.

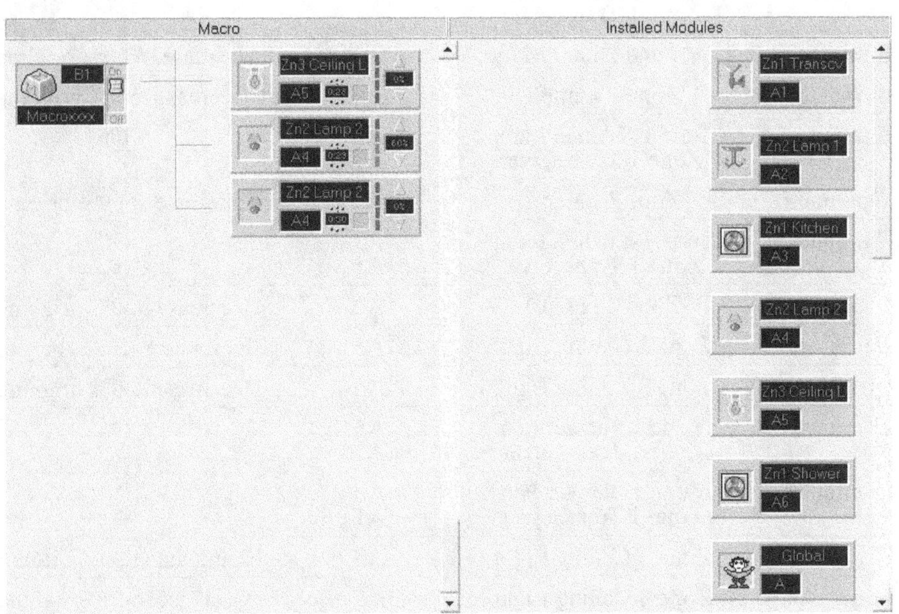

Figure 3-27: Macro Screen After Completing Design Work

99. Make sure that your 3.5" work disk is still in the floppy drive.

100. At the menu bar click on the **File/Save as...** command.

101. In the *Save As* dialog box, select the **a:** drive and type **pgm4xxxx.x10** for the *File name*: entry, where *xxxx* are your initials and those of your lab partner.

102. Save the program, remove your 3.5" floppy disk, and store it in a safe place.

Implementing the Operation of Automated Multizone Residential Power Control Devices

An automated program has been designed at this point, but it has not yet been implemented. It must be run on the Marcraft panel to determine if its operation is accurate. Before doing so, check and correct the steps used to create the automated multizone residential power and control macro, if necessary.

1. Compare the program steps in Table 3-1 with those listed in your macro before initiating its execution on the Marcraft panel.

Table 3-1: Multizone Program Outline

Step	Delay	Module	Address	Operation
1	n/a	Zone 2 Lamp 2	A4	Illuminate at 100% brightness
2	1 minute	Zone 1 Shower Fan Zone 1 Transceiver	A6 A1	On
3	2 minutes	Zone 1 Shower Fan Zone 1 Transceiver	A6 A1	Off
4	3 minutes	Zone 2 Lamp 2	A4	Decrease to 60% brightness
5	4 minutes	Zone 2 Lamp 2	A4	Decrease to 0% brightness
6	5 minutes	Zone 1 Kitchen Lamp Zone 1 Transceiver	A3 A1	On
7	6 minutes	Zone 2 Lamp 1	A2	Illuminate at 60% brightness
8	7 minutes	Zone 1 Kitchen Lamp Zone 1 Transceiver	A3 A1	Off
9	8 minutes	Zone 2 Lamp 1	A2	Decrease to 0% brightness
10	9 minutes	Zone 3 Ceiling Light	A5	Illuminate at 100% brightness
11	10 minutes	Zone 3 Ceiling Light	A5	Decrease to 0% brightness
12	11 minutes	Zone 1 Shower Fan Zone 1 Transceiver	A6 A1	On
13	12 minutes	Zone 1 Shower Fan Zone 1 Transceiver	A6 A1	Off
14	13 minutes	Zone 3 Ceiling Light	A5	Illuminate at 60% brightness
15	14 minutes	Zone 3 Ceiling Light	A5	Decrease to 0% brightness
16	15 minutes	Zone 1 Kitchen Lamp Zone 1 Transceiver	A3 A1	On
17	16 minutes	Zone 1 Shower Fan	A6	On
18	17 minutes	Zone 1 Shower Fan	A6	Off
19	17 minutes	Zone 2 Lamp 1	A2	Illuminate at 100% brightness
20	18 minutes	Zone 1 Kitchen Lamp Zone 1 Transceiver	A3 A1	Off
21	19 minutes	Zone 2 Lamp 1	A2	Decrease to 70% brightness
22	20 minutes	Zone 1 Kitchen Lamp Zone 1 Transceiver	A3 A1	On
23	20 minutes	Zone 2 Lamp 1	A2	Decrease to 60% brightness
24	21 minutes	Zone 3 Ceiling Light	A5	Illuminate at 100% brightness
25	22 minutes	Zone 1 Shower Fan	A6	On
26	23 minutes	Zone 1 Kitchen Lamp	A3	Off
27	24 minutes	Zone 2 Lamp	A2	Decrease to 0% brightness
28	25 minutes	Zone 1 Shower Fan Zone 1 Transceiver	A6 A1	Off
29	26 minutes	Zone 3 Ceiling Light	A5	Decrease to 60% brightness
30	27 minutes	Zone 2 Lamp 2	A4	Illuminate at 100% brightness
31	28 minutes	Zone 3 Ceiling Light	A5	Decrease to 0% brightness
32	29 minutes	Zone 2 Lamp 2	A4	Decrease to 60% brightness
33	30 minutes	Zone 2 Lamp 2	A4	Decrease to 0% brightness

2. When you are satisfied that your macro steps are identical with those listed in Table 3-1, activate the residential power and control system by moving the ac toggle switch to its **ON** position.

3. Make sure that your program macro *pgm4xxxx.x10*, where *xxxx* are your initials and those of your lab partner, is still loaded in the ActiveHome computer program.

4. If necessary, reload it from your 3.5" floppy disk.

5. Click on the **General** folder tab to see the screen shown in Figure 3-28.

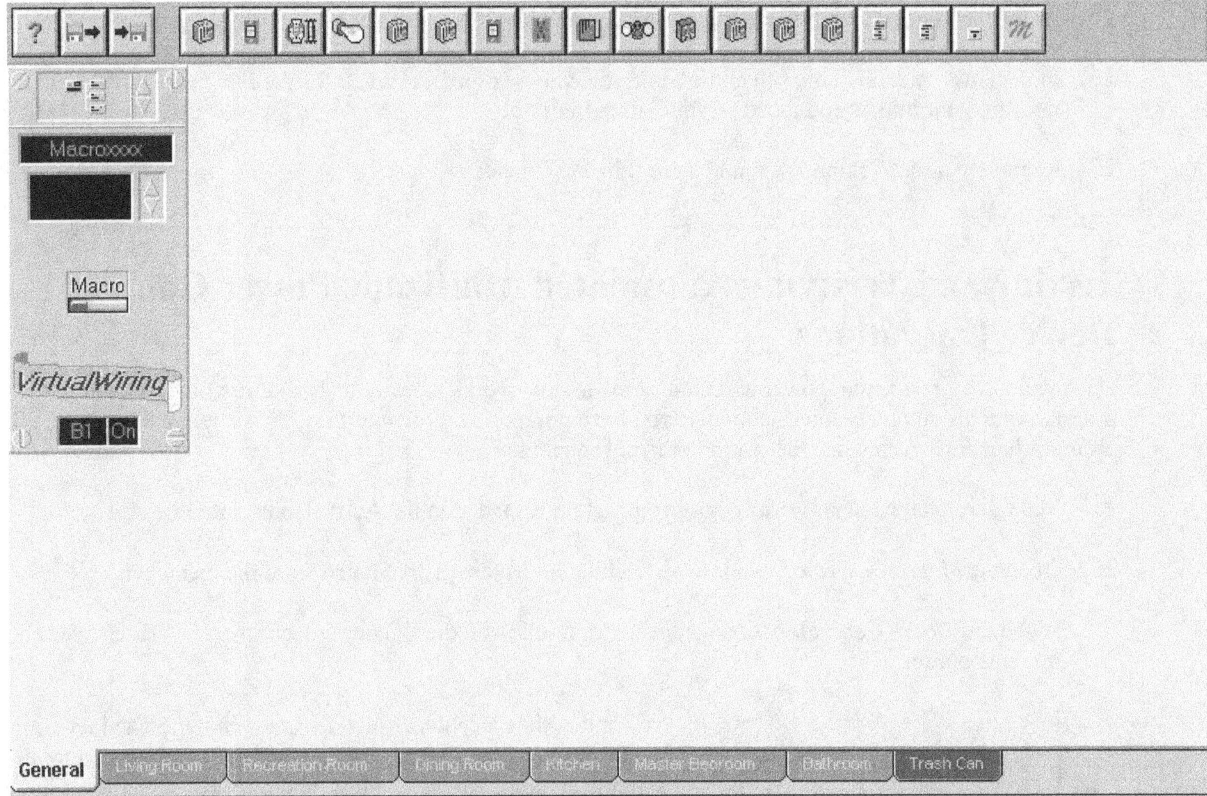

Figure 3-28: Screen View at the General Folder

6. In the switch control area of the visible macro icon, click on the **Macro** button to execute your program.

The macro you created should execute, and the components mounted on the Marcraft panel should react to the commands sent to them by the program.

7. Observe how the various components on the Marcraft panel are reacting to the commands being sent to them.

8. If any commands do not appear to execute as dictated by the program, make a note of them.

9. If all of the commands execute as expected, go to step 14.

10. From the *Macro generator* screen, compare the module steps displayed in the *pgm4xxxx.x10* program with those listed in the outline of Table 3-3.

11. Correct any entries that contain operational data that differs from that shown in Table 3-1.

12. Once your corrections have been entered into the macro, execute the program again.

Looking at the Macro Generator screen at this point. You can always run the cursor over the standard macro placeholder icon, click the *right* mouse button, and then click on the *Test* command.

13. Repeat steps 10 through 12 until all of the programmed commands execute as expected.

14. Make sure that your 3.5" work disk is in the floppy drive.

15. At the menu bar click on the **File/Save as…** command.

16. In the *Save As* dialog box, select the **a:** drive and type **pgm5xxxx.x10** for the File name: entry, where *xxxx* are your initials and those of your lab partner.

17. Remove your 3.5" floppy disk and store it in a safe place.

Testing and Verifying Automated Multizone Power Control Device Operations

Once you have a residential power and control macro that works the way it should, it's time to input the program into the computer interface. After you have programmed the interface, the system will perform all program functions even when the computer is not running.

1. Make sure that the *pgm5xxxx.x10* macro program is loaded in the ActiveHome software.

2. Open the *Macro Generator* window by clicking on **Macro/Edit Macro** from the menu bar.

3. When the *Macro Generator* screen appears, right-click on the standard macro placeholder and select the **Fast** option.

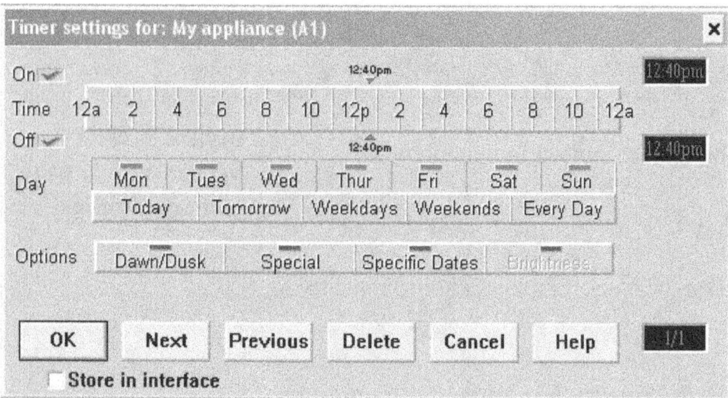

**Figure 3-29:
Timer Settings
Dialog Box**

Notice that the placeholder now has a green button and lettering. Until now, you have been using the standard macro format for use with the computer only. The fast macro format will allow you to upload your macro to the computer interface. Once the program resides on the computer interface, it will run without any further assistance from the computer system.

4. Close the *Macro Generator* screen and return to the *General* folder containing your *Macroxxxx* module icon.

5. Click on the black window area of the on-screen *Macroxxxx* module icon to get to the **Timer settings** dialog box shown in Figure 3-29.

6. Examine the **Timer settings** dialog box carefully.

7. Check to see if the **Store in interface** checkbox has a check in it. If it doesn't, check that box.

8. Click on the green LEDs located above the *Sat* and *Sun* day selector buttons to turn them **OFF**.

This leaves Monday through Friday available as days for which timed events can be activated. For classroom purposes, this arrangement will run an uploaded macro at a specified time, Monday through Friday.

9. Use the mouse pointer to click on the **green on arrow**, and slide it to **10 minutes** from the current time.

For example, if the current time were 9:50 am, you would slide the green arrow to a time of 10:00 am.

10. Click the **Next** button on the *Timer Settings* dialog box. Use the mouse pointer to set the macro to run at least 30 minutes later than the run time you just entered.

For example, if the run time entered previously was 10:00 am, the next run time could not be sooner than 10:30 am. In reality you should allow at least 5 minutes from the end of one program to the beginning of another.

11. Make sure that the macro will run the program from the hours of **8:00 am to 4:00 pm**, **Monday through Friday**, without computer assistance.

The *Timer settings* dialog box permits you to run the macro at least four times. Turning OFF the power at the panel can always interrupt the activity when operation is not desired. As you can see, the programming permitted by the *Timer settings* dialog box is much more sophisticated than what we are doing. In fact, certain dates can be specified without regard to any other settings.

12. When you have finished specifying the run times for your macro, click on the **OK** button.

The *General* folder screen now appears as shown in Figure 3-30, with the earliest ON time displayed in the *Macroxxxx* module's black *Control Time On* window.

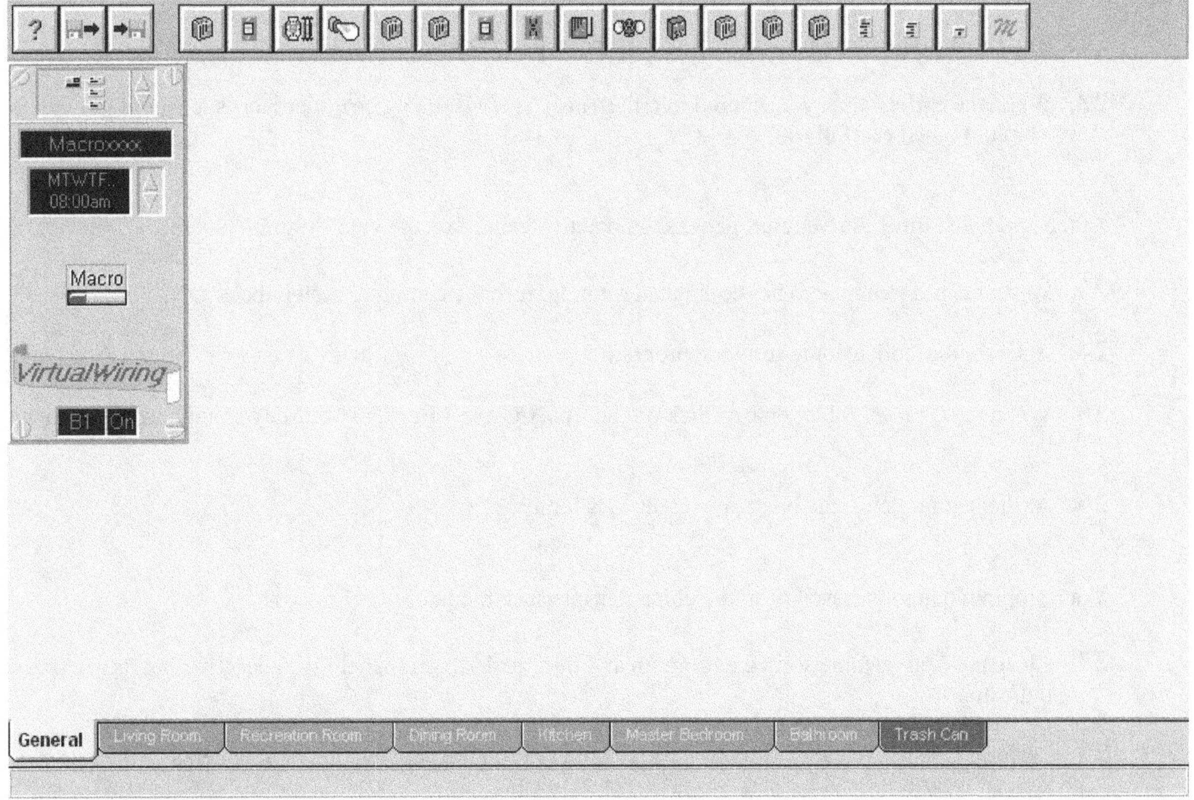

Figure 3-30: Screen View of Timer Settings at the General Folder

13. Click on the **UP arrow** to the right of the **Control Time On** window to check the accuracy of the additional ON time you specified in the *Timer settings* dialog box.

14. If any ON time is not correct, make the necessary corrections before the next step.

15. Make sure the Marcraft panel is powered and the red power indicator is activated.

16. From the menu bar click on **Tools/Download Timers and Macros**.

As you download your macro to the computer interface, the progress of the operation is displayed as shown in Figure 3-31.

Figure 3-31: Screen View of Macro Downloading Procedure

17. Make sure that your 3.5" work disk is in the floppy drive.

18. At the menu bar click on the **File/Save as...** command.

19. In the *Save As* dialog box, select the **a:** drive and type **pgm6xxxx.x10** for the *File name:* entry, where *xxxx* are your initials and those of your lab partner.

20. Remove your 3.5" floppy disk and store it in a safe place.

21. Exit the ActiveHome program and turn the computer **OFF**.

22. When the earliest time you entered into the *Timer settings* dialog box approaches, observe the Marcraft panel carefully.

At the appointed time, the Marcraft panel should activate and execute your program macro.

23. When the program macro has completed running, turn the computer system back **ON**.

24. Activate the ActiveHome software program.

25. When the *General* folder opens, click on the **Tools/Clear Interface Memory** command on the menu bar.

26. At the warning message box click on the **OK** button.

Your program macro is erased from the computer interface module.

27. Close the ActiveHome software program and turn the Marcraft panel **OFF**, deactivating its red power indicator.

28. Perform a general cleanup of your lab area, and return all equipment and components to their proper storage locations.

LAB QUESTIONS

Feedback

1. What became of the "Family Room" sheet tab in the ActiveHome desktop?

2. What happened to all the module icons that were originally located in the General tab folder?

3. To which sheet layout was the "Zn2 Lamp 1" module icon located?

4. What is the shortest time delay that can be programmed for any module using the ActiveHome software?

5. Why does the nightlight plugged into the transceiver activate simultaneously with any fan?

6. How can a macro program be tested when looking at the Macro Generator screen?

7. How long does the macro program in this lab procedure require to complete?

8. How is the memory on the computer interface module erased?

Residential Power and Control Systems Research

OBJECTIVES

1. Search for external power control systems.
2. Research various X-10 power-line components.
3. Locate applicable local, state, and federal regulations.

Power and Control Systems

RESOURCES

1. Books
2. Newspapers
3. Magazines
4. Computer with Internet access and graphic capture software

DISCUSSION

The ability to conduct serious research is a skill, and a most underrated one at that. Successful scholarship is less daunting than it once was, due to the many sources of information now available to the modern student. But traditional researching methods still require a genuine effort. This includes the careful examination of library card files, along with serious reading of books, newspapers, and magazines. With the advent of the Internet, additional opportunities for conducting information searches have arisen. Because of this, modern libraries encourage Internet research by providing computers with Web access.

Unfortunately, inquisitive minds run the risk of drowning in an ocean of unwanted information by using Web-based search engines before mastering basic researching skills. Becoming ensnared in time-wasting searches through mountains of available data hinders the serious researcher from quickly locating useful information. This situation arises because items of online information items are not readily catalogued for identification and retrieval. Public libraries, as big as some of them are, can still keep ahead of the curve by keeping materials well organized. The billions of individual files residing on the Internet are not so easily manipulated. In the time taken to read this paragraph, thousands of additional files have been added to the Internet. Others have been renamed, relocated, or simply removed. When conducting computer research, the student must know which data delivery service to select from the many available. These include electronic mailings, file transfers, interest group memberships, interactive collaborations, and multimedia displays.

Web browsers are equipped with tools that effectively handle all of the data services previously mentioned. Yet, although web research offers access to much of what is available on the Internet, one major weakness in using it has already been alluded to. The addresses of Internet sites (the information locations) frequently change, or the sites disappear altogether. It's as if a book you were able to check out of the library last month, is no longer available this month! When the recorded Web address for the desired information is entered on the address line and executed, a screen similar to the one shown in Figure 4-1 may appear. Clearly, the information stability we have grown to expect at the public library does not exist on the Internet!

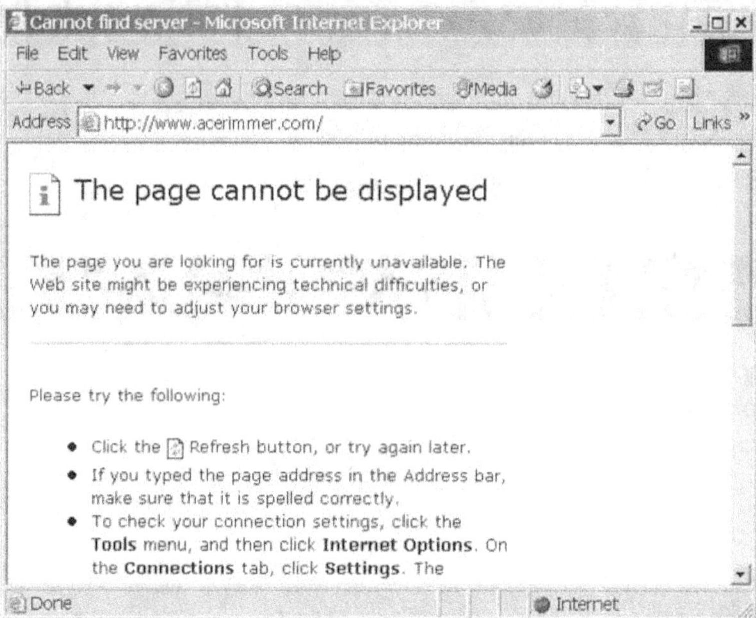

Figure 4-1: Internet Site or Page Cannot be Displayed

One method of conducting Internet research is to copy and store any information (text and graphics) that appears useful to the project at hand, when possible. If the Internet site from which the material was discovered disappears, the information itself is still available via the saved files. Keep in mind that certain Internet sites contain information that is impossible to transfer directly to your computer, especially product vendor pages containing proprietary information, or from information services that require some sort of paid membership. In these cases, you may have to type the relevant text, by hand, to a word processor for data storage, or to capture a necessary graphic using a third-party graphics tool that can convert it into a specified picture format.

It is important to verify the information you obtain by research, especially on the Internet. The appearance of interesting tidbits on an Internet site, or in a newspaper, magazine, or book article, does not guarantee their accuracy or integrity. Their veracity can be strengthened considerably by turning up identical information from a variety of sources!

PROCEDURES

Power and Control Systems

Because external power control systems are becoming increasingly important to residential clients, your beginning research should focus on identifying those systems, and learning how they augment the workings of a modern automated home environment.

1. Examine the following list of external power control systems and components for residential application. The RESI Environmental Control technician will need to be familiar with the items shown here.

- External thermostat controllers
- Hydro power systems
- Radio-controlled ac shutoffs
- Residential power generators

- Service disconnects
- Solar power systems
- Utility meters
- Wind turbine systems

2. Use Tables 4-1 through 4-8 to organize the specified details about the external power control systems and components listed above. For each item, try to locate at least two vendors.

Table 4-1: External Thermostat Controllers

Model	Description	Vendor	Cost

Table 4-2: Hydro Power Systems

Model	Description	Vendor	Cost

Table 4-3: Radio-Controlled ac Shutoffs

Model	Description	Vendor	Cost

Table 4-4: Residential Power Generators

Model	Description	Vendor	Cost

Table 4-5: Service Disconnects

Model	Description	Vendor	Cost

Table 4-6: Solar Power Systems

Model	Description	Vendor	Cost

Table 4-7: Utility Meters

Model	Description	Vendor	Cost

Table 4-8: Wind Turbine Systems

Model	Description	Vendor	Cost

Another concern for the RESI Environmental Control technician is the various components that make up the residential power and control system. Because of the many types of products and components available, it is important to select components that will work together as a complete system. To make your remaining research time more productive, concentrate on information about X-10 power and control components in the product areas listed below. This research task will be of benefit when weighing options for a residential power and control system installation, including its size, layout, and the particular requirements of the client. Information gathered here will help when considering and comparing specific pieces of residential power and control equipment, their system capabilities, available vendors, and estimated costs.

3. Examine the following list. It identifies various X-10 residential power and control components with which an RESI Environmental Control technician will most likely be working.

- Computer controllers
- Dry contact sensors
- Humidity sensors
- HVAC thermostats
- In-line dimmers
- Lamp modules
- Light sensors
- Motion detectors
- Noise filters

- Plug-in dimmers
- Programmable controllers
- Surge protectors
- Telephone responders
- Thermostat controllers
- Timers
- Wall receptacles
- Wall switches
- Wireless remotes

4. Use Tables 4-9 through 4-26 to organize the specified details about the X-10 residential power and control components listed above. For each item, try to locate at least two vendors.

Table 4-9: Computer Controllers

Model	Description	Vendor	Cost

Table 33-10: Dry Contact Sensors

Model	Description	Vendor	Cost

Table 33-11: Humidity Sensors

Model	Description	Vendor	Cost

Table 33-12: HVAC Thermostats

Model	Description	Contract	Cost

Table 33-13: In-line Dimmers

Model	Description	Contract	Cost

Table 33-14: Lamp Modules

Model	Description	Contract	Cost

Table 33-15: Light Sensors

Model	Description	Contract	Cost

Table 33-16: Motion Detectors

Model	Description	Contract	Cost

Table 33-17: Noise Filters

Model	Description	Contract	Cost

Table 33-18: Plug-in Dimmers

Model	Description	Contract	Cost

Table 33-19: Programmable Controllers

Model	Description	Contract	Cost

Table 33-20: Surge Protectors

Model	Description	Contract	Cost

Table 33-21: Telephone Responders

Model	Description	Contract	Cost

Table 33-22: Thermostat Controllers

Model	Description	Contract	Cost

Table 33-23: Timers

Model	Description	Contract	Cost

Table 33-24: Wall Receptacles

Model	Description	Contract	Cost

Table 33-25: Wall Switches

Model	Description	Contract	Cost

Table 33-26: Wireless Remotes

Model	Description	Contract	Cost

Applicable Regulations

A completed residential power and control installation may indeed testify to the skill and awareness of the RESI Environmental Control ttechnician who installed it. Conversely, failure to consult the local, state, or federal bodies having jurisdiction over residential power and control systems may reveal weakness in his/her work ethic.

Local

Strict adherence to federal and state regulations will often result in totally compliant residential power and control systems, but this does not guarantee an automatic free pass by the local coding authority. Special local power use ordinances must also be taken into account before any installation begins! These codes most likely involve local governmental power controlling structures rather than residential installations, but not always. The advent of wireless power control equipment may have resulted in certain city or county restrictions.

The operation of electrical equipment or transmission devices in a manner that habitually or detrimentally affects or interferes with electrical service in a residential neighborhood is illegal in most jurisdictions. The RESI Environmental Control technician should be aware of local codes so that any situation likely to cause this type of trouble can be brought to the client's attention prior to installation.

The local planning commissions or coding departments are good sources for this type of information. They usually control the use of electrical power in the local community including wireless controllers. For requirements concerning power controlling installations, equipment, and uses not specifically addressed by state or federal codes, contact your local planning commission. These guidelines may be enforced by either city or county agencies. For example, the location of various buildings or structures could place unique restrictions on the use of certain wireless power controlling equipment types. The operation of certain types of transmission equipment may be restricted in or near hospital zones.

1. Use Table 33-27 to organize specific details about the power and control systems and equipment strictly regulated by your local codes. Try to find more than one source of information for each consideration or restriction.

Table 33-27: Local Power and Control System Codes or Regulations

Information Source	Specified Equipment	Restrictions	Special Considerations

State

State-sponsored power and control system regulations normally center on providing continued power to state government buildings, and avoiding interference to power controlling equipment from either direct sabotage or inadvertent wireless transmissions. State agencies also seek to avoid inadvertent interference with residential power service from state-sponsored power controlling equipment. State universities publish their own policies regarding the legitimate use of their power and control systems. As might be expected, these policies normally restrict usage to supporting departmental mandates and conducting legitimate government or college business.

State utility regulators investigate specific statewide power failures, especially severe ones, in order to determine the exact causes. These investigations may extend to systems that lie outside state or national boundaries, but are nevertheless accessed via the existing international power grids. Consequently, energy market forces in the United States are currently restructuring, with each state in various stages of development. Accordingly, various energy regulations and policies are being adopted, with the end goal being electric power deregulation. This deregulation seeks to split the electric industry into two parts, a supplier side and a distributor side. The electric generation companies will manage the supply side, while the transmission, distribution, and wire companies will run the distribution side.

The need for electric power that is often generated great distances from where it is needed has motivated this push towards deregulation. The resulting policies and regulations are connecting Distributed Generation (DG) systems to the grid. Distributed generation brings the electricity supply apparatus closer to the point of use, substantially reducing grid congestion. This idea is especially important in areas where power is generated locally. These locations often experience high demand from remote users, with the distribution systems already operating at or near capacity. Another benefit of distributed generation is the reduction of the line losses associated with transmitting electricity long distances. Information concerning the ongoing deregulation of the electric industry, by your own state government can be found by browsing its Web pages.

2. Search the information pages of your state government and locate the applicable energy policies regarding its electric generation and distribution systems. Indicate whether the policy status is current, in the process of being adopted, and directed towards a DG system. Organize the specific details in Table 33-28.

Table 33-28: State Electric Generation and Distribution Policy

State Agency Responsible	Applicable Policy	Applicable Policy Status	DG System Goal

3. Research your state's information pages to find any power generation companies located within state boundaries, and determine whether these companies operate DG systems, operate as remote electric suppliers, or can be considered to be in both categories. Do they operate at peak capacity, or less? Do they provide reliable service, or do their facilities need an overhaul? Organize their specific details in Table 33-29.

Table 33-29: State Electric Generation Company Details

State Power Facility	Remote or DG System	Current Operational Capacity	Quality Service Provided

Federal

The explosion of technical development during the last several years makes it more important than ever to officially determine the suitability of one product over another. In order to attain some control over this activity, various technical bodies generally recognized by manufacturers and consumers around the world have been vested with the authority to make this determination.

Consult the major organizations discussed here regarding the standards that determine the acceptability of both products and labor involved in power and control system installations and operation. The RESI Environmental Control technician can depend on the standards published by these organizations to guide him or her in the selection of appropriate equipment for a particular installation.

National Electrical Code

The National Electrical Code contains information upon which all aspiring electricians must be tested before obtaining their licenses. It is sponsored by the National Fire Protection Agency, and most electricians can easily locate any required information in the code. Keep in mind, however, that the National Electrical Code is merely a guideline. Most states require a permit and an inspection for performing any electrical work. As comprehensive as it is, the NEC alone cannot guarantee safe electrical installations. This is why an RESI Environmental Control technician should use his or her own judgment when undertaking electrical work, yet exceed the minimum safety standards set forth by the NEC.

You should contact the local city or town wire inspector before wiring any power and control system, because each state may differ slightly in its requirements for inspection and code compliance. The local wire inspector is responsible for rule interpretation and for code enforcement, being "The Authority Having Jurisdiction" in most locations.

4. Identify the local wire inspector in your area. Record the information specified in Table 33-30 to organize the details about your local wire inspector.

Table 33-30: The Authority Having Jurisdiction

Business Name	Personal Name	Street Address	Telephone Number	Web Address

TIA/EIA Standards

Telecommunications Systems Bulletins (TSBs) are addenda to, or explanatory comments about, either an industry standard or an interim standard, and are often integrated into the next standard revision. The standards themselves, submitted by the Telecommunications Industry Association (TIA) and the Electronic Industries Alliance (EIA), are certified by the American National Standards Institute (ANSI).

5. Conduct an Internet search for any ANSI/TIA/EIA standards pertinent to residential power and control systems, and record information about them in Table 33-31.

Table 33-31: ANSI/TIA/EIA Residential Power and Control System Standards

Standard ID	Standard Name	Standard Description	Year Established

IEEE Standards

The IEEE is important because of the size and scope of its membership. Consisting of more than 377,000 individual members from more than 150 different countries, the IEEE's full name is the Institute of Electrical and Electronics Engineers, Inc. A nonprofit, technical professional association, the IEEE is a leading authority in technical areas ranging from computer engineering, biomedical technology, telecommunications, electric power, aerospace, and consumer electronics.

6. Conduct an Internet search for any IEEE standards pertinent to residential power and control systems, and record information about them in Table 33-32.

Table 33-32: IEEE Residential Power and Control System Standards

Standard ID	Standard Name	Standard Description	Year Established

Underwriters Laboratories Inc. (UL)

Founded in 1894 as an independent, not-for-profit, product safety testing and certification organization, Underwriters Laboratories, Inc. has tested products for public safety for more than 100 years. UL holds an undisputed reputation as the most recognized and reputable certification, conformity assessment, and product safety provider in the world, not just the United States.

UL lists various manufacturers for specific types of installations and equipment, tests the equipment for specific uses, and judges installations against the governing standards. At the time of this publishing, UL services also include helping companies achieve global acceptance for their electrical devices, programmable systems, and quality processes.

7. Using an Internet search, locate several UL standards that have been incorporated by their manufacturers into some of the power and control components listed earlier. Record the information you find about applicable UL standards in Table 33-33. Feel free to extend the table if necessary.

Table 33-33: UL Power and Control Component Standards Incorporated by Manufacturers

Power and Control Component	Manufacturer	Standard ID	Specified Requirements

National Fire Protection Association (NFPA)

Since its founding in 1896, the National Fire Protection Association (NFPA) has developed and advocated a scientifically based consensus of codes, standards, research, training, and education. It has pioneered in the effort to reduce the worldwide burden of fire and other hazards on quality of life.

The NFPA's headquarters is located in Quincy, Massachusetts. Consisting of more than 75,000 members representing nearly 100 nations, and 320 employees around the world, the NFPA is an international, nonprofit membership organization that sets installation guidelines for fire equipment, and is the world's leading advocate of fire prevention. It is an authoritative public safety source and holds classes to instruct various installers on how to follow the guidelines.

8. Using an Internet search, locate several NFPA standards that have been incorporated by their manufacturers into some of the power and control components listed earlier. Record the information you find about applicable NFPA standards in Table 33-34. Feel free to extend the table if necessary.

Table 33-34: NFPA Power and Control Component Standards Incorporated by Manufacturers

Power and Control Component	Manufacturer	Standard ID	Specified Requirements

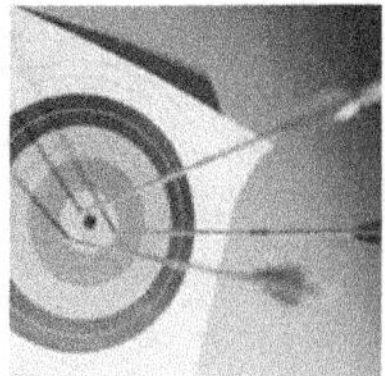

LAB QUESTIONS

Feedback

1. Why can a utility meter be considered to be an external power control system?

2. What is a radio-controlled ac shutoff?

3. What are several uses for a motion detector?

4. What is an X-10 telephone responder used for?

5. What does the acronym DG stand for?

6. How will deregulation affect the electric power industry?

Whole-House Lighting Control and Management Systems

OBJECTIVES

**Power and
Control Systems**

1. For the given scenario, select a cost-effective lighting control and management system that is sized to meet the customer's requirements.
2. Identify the modules and interconnection technologies needed to meet the customer's requirements.
3. Configure the new system.
4. Apply standard numbering to the lighting control system.
5. Explain how you plan to measure the load requirements of the new system.
6. Define the types of hardware modules that you will purchase for the new system.
7. Define the functions of the lighting control unit.
8. Plan a wiring topology for the system.
9. Decide on actions that would improve the security aspects of the design.
10. Plan a wiring topology for the lighting system in the home theater room.
11. Explain the benefit of interfacing the smoke alarm system with the lighting circuits.
12. Recommend an alternative to large banks of light switches.
13. Outline switch configuration options.
14. Determine the cost of the X-10 network components required for this project.
15. Provide advice on the control of a single light from two separate locations.
16. Recommend appropriate installation guidelines for the electrical contractor.
17. Define a cable type for the landscape lighting system.
18. Define a cable type for use in conduits.
19. Define basic procedures for troubleshooting a lighting circuit.

RESOURCES

1. Marcraft's *RESI Environmental Control Endorsement* textbook
2. Pencil

DISCUSSION

Read the following scenario:

Smart Automation Systems (SAS) Ltd. is a home control systems integrator, based in Tucson, Arizona. The firm specializes in designing, installing, and servicing integrated lighting control and management systems.

The firm has recently teamed up with a local interior designer on a new project that involves the installation of a sophisticated home electronics system.

You are an independent systems integrator who has collaborated previously with SAS on numerous residential installations. You have now been subcontracted by SAS to manage the planning and installation of the lighting control element of the overall project. Completion of the overall project is expected to take twelve months.

Table 5-1 includes specifications and requirements gathered from interviews and written documentation.

Table 5-1: Specifications and Client Requirements

Requirement Category	Details
Aesthetics	The lighting management technology must fit with the style and décor of the completed house. All controlled outlets will be permanently enclosed in electrical connection boxes.
Control	The new system will provide simple, convenient control of all home lighting. All switches and outlets will be controlled by the new system.
Remote access	Your client wants to use a standard touchtone phone to access the lighting control system.
Outdoor lighting	In addition to the main lights for the driveway and back yard, your client has also requested the installation of a landscape lighting system.
Security	The home security and smoke detection systems will be integrated into the lighting control system.
Zoning	Each room on the ground floor of the house will represent a zone in the new lighting control system. The lighting for the bedrooms and bathrooms will share one zone. The last zone will be configured for the hallway and stairs.

Power and Control Systems

PROCEDURES

Planning and Design

1. Note each zone on the lighting plan in Figure 5-1. Use zone codes with a standard numbering scheme to label each element of the lighting control system on the upper floor of the house.

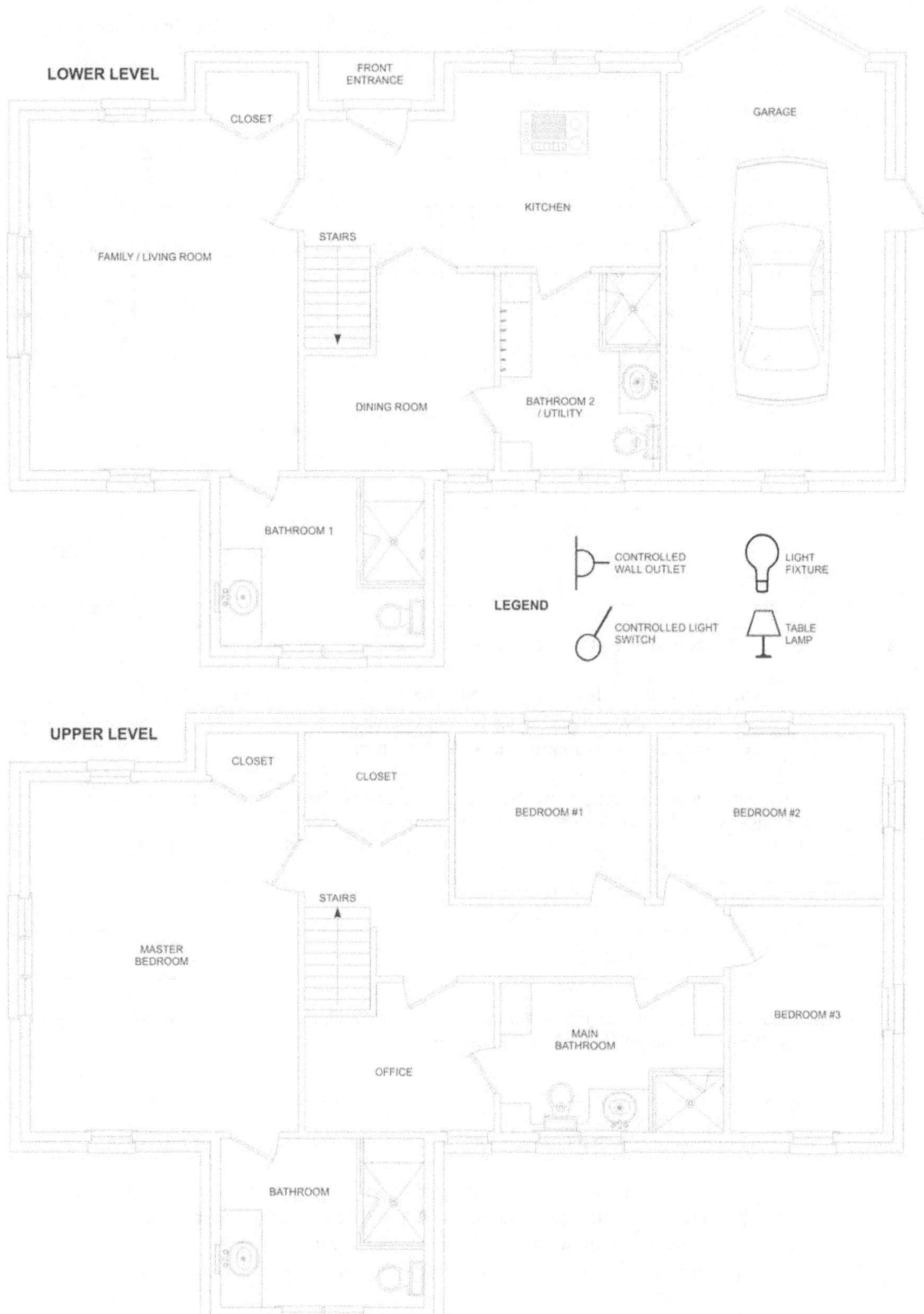

Figure 5-1: Drawing to Label the Floor Plan of the House

2. How do you plan to calculate the estimated levels of power that the home lighting system will consume?

____

____

____

3. The outlets, switches, and the central controller for the new home lighting control and management system will be based on which of the following protocols?

    ____ a. TCP/IP
    ____ b. HomePNA
    ____ c. HomeRF
    ____ d. X-10

Explain your answer:

____

____

____

4. The new X-10 transmitter (controller) unit will be used:

    ____ a. to link the lighting system to the home's security system.
    ____ b. to send low-voltage signals over existing wiring.
    ____ c. to monitor and control access to the Internet.

5. After reviewing the design requirements, you have to choose between two wiring topologies – home run and daisy chain. Explain where each wiring topology is used.

____

____

____

6. Numerous technologies are available for controlling lights other than manual control from light switches. Which one do you plan to install, and why?

____

7. Which of the following design elements would improve the security aspects of the design?

    ____ a. Integrate motion sensors with the outdoor lighting system
    ____ b. Integrate photoelectric sensors with the outdoor lighting system
    ____ c. Integrate inertia sensors with the outdoor lighting system
    ____ d. Integrate ultrasonic sensors with the outdoor lighting system

8. The lighting control system for the home theater is extremely important for creating an ambient atmosphere for your client's family. The layout of the home theater shown in Figure 5-2 includes four separate lights that use X-10 receiver modules to dim the lights. On Figure 5-2 draw the wiring scheme that will be used to connect the four lights back to the central controller. Explain your reasons for choosing this wiring scheme.

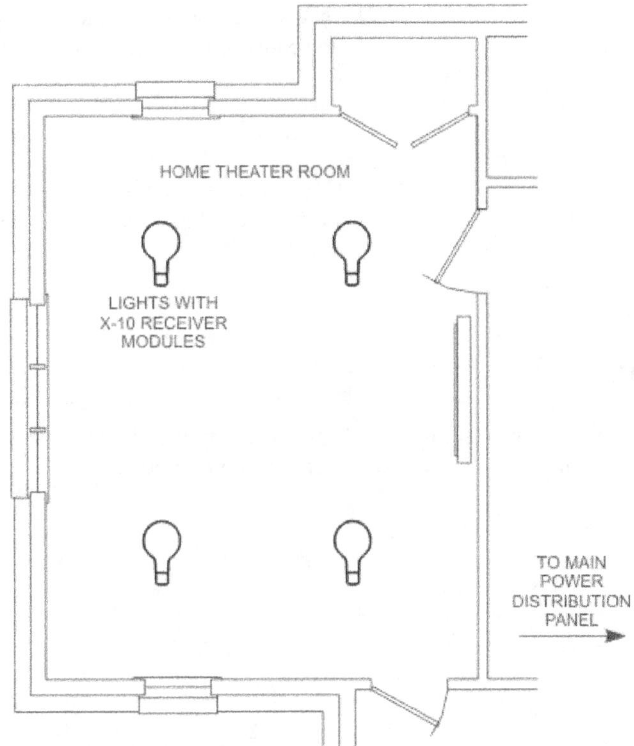

**Figure 5-2:
Drawing to Label
the Home Theater**

9. One requirement of this project is to wire the smoke detection system into the lighting circuits. How will this benefit your client?

10. Your client is reluctant to install large banks of light switches. What piece of hardware should you install as an alternative to switches?

Wall switches are the most popular switch used for residential light control. The types of wall switches available vary according to where they are installed, the type of circuit they are controlling, and the current rating of the load connected to them. In this project you will inform your client of the available switch options.

11. In Table 5-2 match each switch configuration description with its corresponding block abbreviation and nomenclature.

Table 5-2: Switch Configuration Descriptions

Abbreviation and Nomenclature	Switch Configuration Description
	This type of switch is used to operate a single light fixture.
	This type of switch is referred to as a "3-way" and is used to control a light from two locations.
	This type of switch is an "on-off" type control, and is used to interrupt a 2-wire circuit.
	This type of switch is a "changeover" type that is normally used to control 220-volt, 2-phase circuits.

Project Quotation

1. On the bill of materials form shown in Figure 5-3, list the components required for this project. Prices may be obtained from an online supplier of home networking components.

Figure 5-3: Bill of Materials Form

Customer Name: _______________________________________

Address: ____

Project ID No: ____

Description of Component	Catalog Number	Number Required	Cost Per Unit	Total Price
Total Cost of lighting control components			$	

System Installation and Setup

1. How does your client benefit from an option to control the hall light from the top and bottom of the stairway?

____

2. You will recommend appropriate guidelines for the electrical contractor responsible for installing the new system. In Table 5-3, identify which are acceptable installation procedures (yes), and which are unacceptable (no). If the procedure is unacceptable as written, explain what is acceptable.
For security purposes your client wants to install a switch that activates the outdoor lighting system when the light intensity, or darkness, reaches a certain value. What type of switch should you install to fulfill this request?

Table 5-3: Appropriate Guidelines

Installation Guidelines	Yes	No	Reasons
For locating the outlets for lighting fixtures, duplex receptacles are normally placed six to eight feet apart in newer homes.			
The lighting modules for all the bedrooms and the kitchen should be on one circuit.			
22 AWG (American Wire Gauge) wire, with ground, should be used to wire most elements of the new lighting system.			
GFCI protection should be provided only in the kitchen and the main bedroom.			

____

____

3. To create a warm and welcoming atmosphere for the exterior of the new house, your client has requested a landscape lighting system. Which of the following types of cable should you use for this system?

   ____a. Low-voltage wiring
   ____b. NM cable
   ____c. MC cable
   ____d. AC cable

Why? ____

____

____

4. Sections of the cabling for the lighting system will be run through conduit, which could lead to some abrasion damage. Which of the following NEC-designated cable types should be used to minimize the abrasion?

   ____a. T
   ____b. TW
   ____c. HH
   ____d. THWN

Explain your answer:

____

____

____.

Troubleshooting

The final phase in the implementation of a fully operational home lighting system is testing and troubleshooting.

1. In Table 5-4, put the procedures for troubleshooting the home lighting system in the correct sequence.

Table 5-4: Home Lighting System Troubleshooting Sequence

Sequence No.	Procedure
	Place the tester on the wiring, or in the slots of the ac outlet.
	Turn off the circuit breakers, or remove the fuse that controls power to the circuit you are working on.

Structured Wiring

Power and Control Systems

OBJECTIVES

1. Design a cost-efficient home wiring solution that will support existing and future needs.
2. Purchase all necessary equipment and supplies.
3. Calculate the number of home runs that complies with international cabling standards.
4. Choose suitable types of cable for the installation.
5. Select a cable type for the video distribution system.
6. Suggest a location for the distribution center.
7. Match the cable type to a corresponding home technology subsystem.
8. Define the cabling requirements for the surround sound system.
9. List the actions required to install the new wiring system.
10. List acceptable installation procedures.
11. Design a labeling system.
12. Identify the standard to be used when installing this system.
13. Formalize a termination standard for RJ-45 connections.
14. Select a device to distribute audio and video throughout the house.

RESOURCES

1. Marcraft's *RESI Environmental Control Endorsement* textbook
2. Pencil

DISCUSSION

Read the following scenario:

Smart Homebuilders in Colorado wishes to provide all their customers with structured cabling as a standard amenity. The company is scheduled to complete a development of fifty new homes this year and has commissioned you to manage the installation of the structured wiring systems.

Table 5-1 includes general requirements that were gathered from interviews and written documentation.

Table 6-1: Structured Wiring Requirements

Requirement Category	Details
Functionality	The design must provide home owners with a range of services including multiroom audio, Internet access throughout the house, TV distribution, and telecommunication services to all rooms.
Project Estimating and Procurement	Material and labor for the wiring system must cost less than $1,800 per house.

PROCEDURES

Preproject Planning

1. The ANSI/TIA/EIA 570A Residential Telecommunications Cabling Standard establishes minimum
    "grades" of residential cabling, and is generally accepted as the minimum requirement for the design
    of residential low-voltage structured wiring. Based on the information contained in Figure 6-1, what
    is the total number of home wire runs required to comply with this standard?

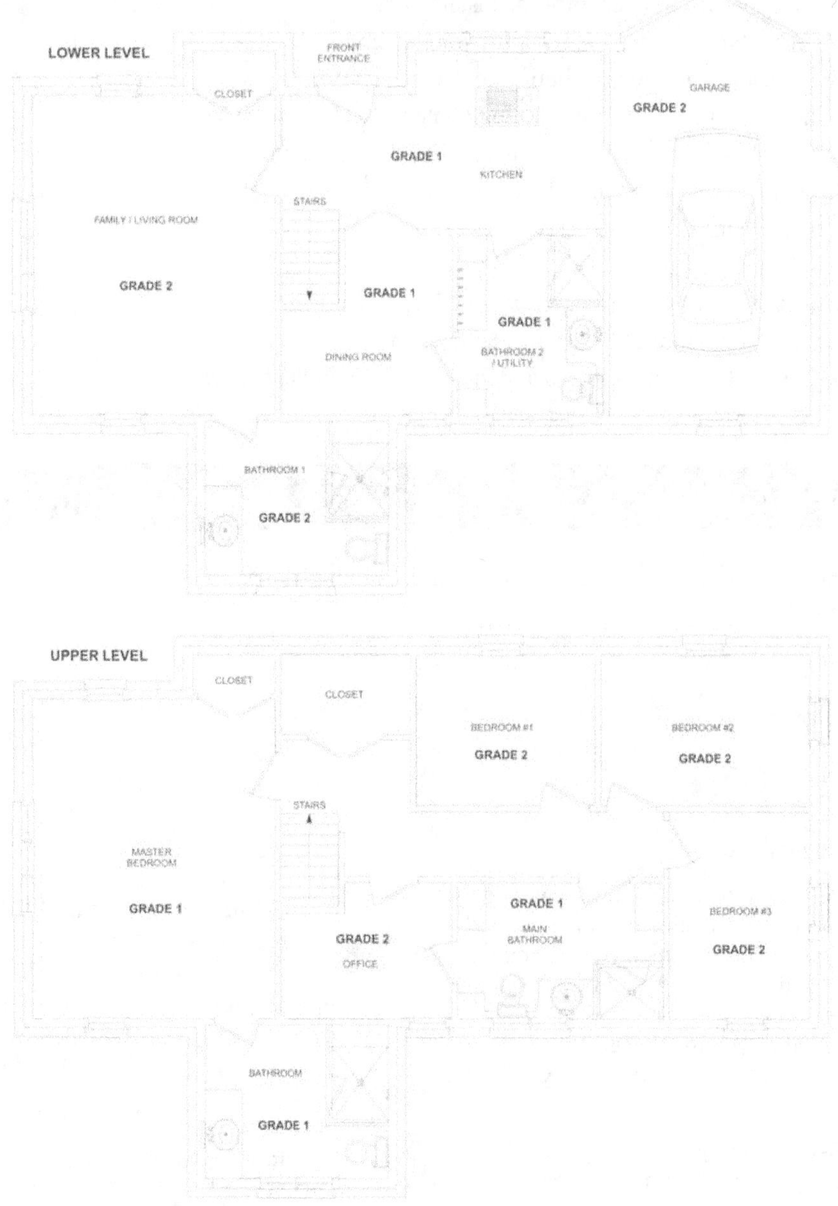

**Figure 6-1:
Home Wiring Plan**

2. Choosing a wiring system is one of the most important decisions of the overall project. After reviewing the design requirements, you have to choose between four types of systems. Which of the following cabling systems are appropriate for this design? (Select all correct answers.)

    ____a. Fiber

    ____b. UTP cable

    ____c. Coaxial

    ____d. Shielded twisted pair

3. UTP cable is graded and tested according to a set of performance criteria. The grading system divides UTP cable into separate categories. Match each category type to the characteristics in Table 6-2.

- Category 1

- Category 2

- Category 3

- Category 4

- Category 5

- Category 5e

Table 6-2: Structured Wiring Requirements

Category	Cable Characteristics
	This category carries a rated bandwidth of 100 MHz, and is widely used to support Fast Ethernet networks.
	This category has a rated bandwidth of 20 MHz, and was seldom used outside of IBM Token Ring LANs.
	Designed for transmission speeds of up to 1 gigabit per second.
	Originally used to support Ethernet and Token Ring networks. It has a rated bandwidth up to 16 MHz.
	A common voice-grade telephone cable, this category is rarely used in modern residential structured wiring installations.
	With a rating of only 1 MHz, this category of cable is now obsolete.

4. Choosing the correct type of cable to distribute the video signal throughout the house is an important decision. After reviewing the design requirements, you have to choose between three types of coaxial cables currently available — RG-59, double RG-6, and Quad RG-6. Which one will you install?

____

Why?

5. The home will include different types of audio and video equipment. How might you "future proof" the wiring infrastructure used to interconnect these devices?

____

____

6. On Figure 6-2, draw and label the proposed location of the centralized distribution box.

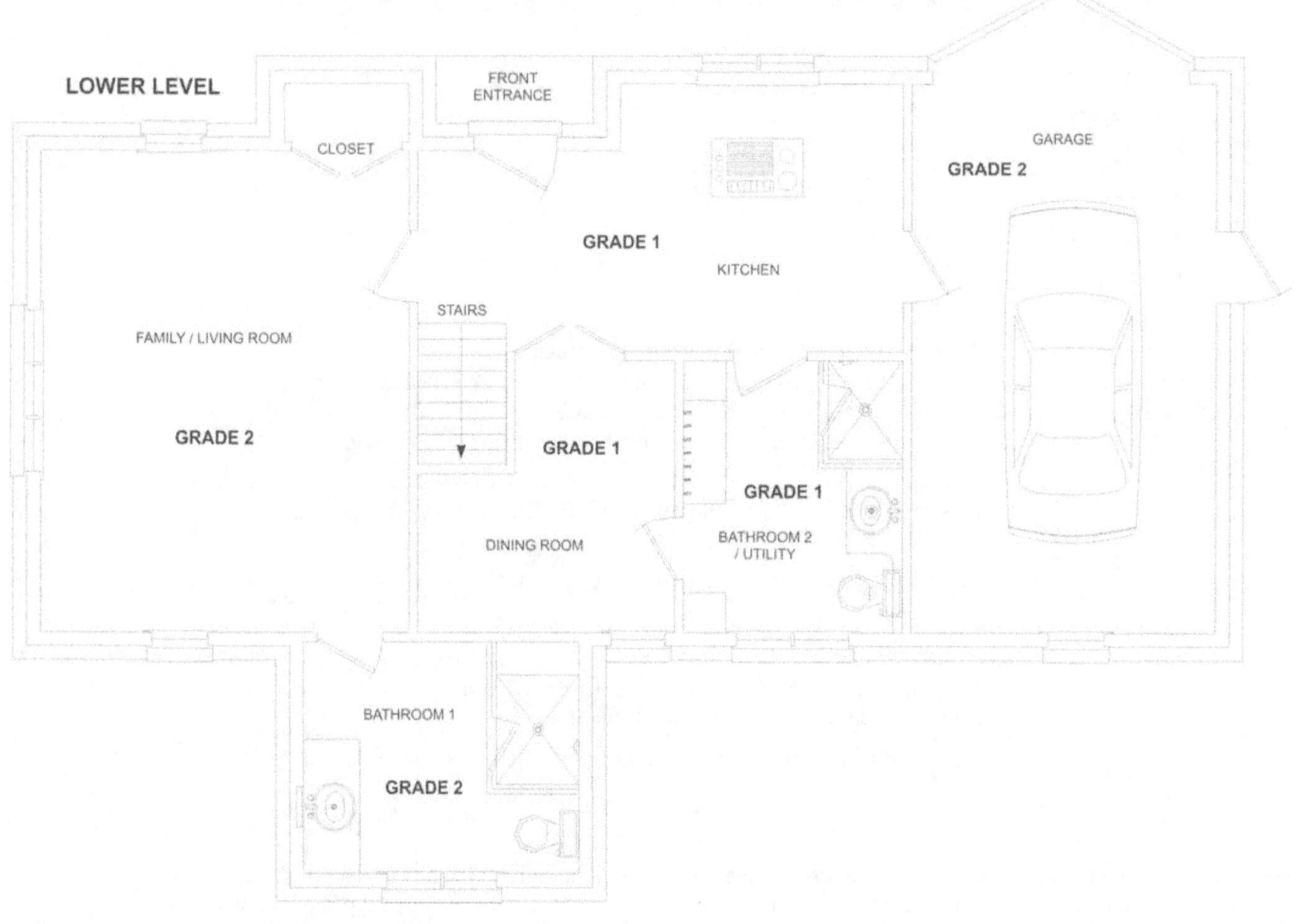

Figure 6-2:
Label Location of
the Centralized
Distribution Box

Explain your reasoning:

____

____

Device Connectivity

1. In Table 6-3, match each home technology subsystem to a corresponding cable type.

- Data and telecommunications
- TV
- Security
- Backbone

Table 6-3: Home Technology Subsystems

Type of Cable	Technology Subsystem
Category 5	
Fiber	
Coaxial	
Category 3	

2. The residential floor plans for this project include space for a home theater. What are the cabling requirements of the surround sound system?

____

____

3. In Table 6-4, list the actions needed at each stage of the installation to fulfill the requirements of this project.

Table 6-4: Home Technology Installation Actions

Implementation Stage	Actions
Prewire (Rough-in)	
Trim-out	

4. In Table 6-5 identify which of the following procedures are acceptable installation procedures (yes) and which are unacceptable (no). If the procedure is unacceptable as written, explain what is acceptable.

Table 6-5: Acceptable Installation Procedures

Procedure	Yes	No	Acceptable Procedure
Drill holes at the same height as electrical outlets.			
Run Category 5 cable wire parallel to power wiring.			
Video wiring is usually terminated with a RCA connector on each wire.			
The bend radius of a 4-pair UTP cable should not exceed two times the cable diameter.			

5. The wire behind each outlet needs to be clearly labeled. On Figure 6-3 use a pen or pencil to assign a label to every cable run.

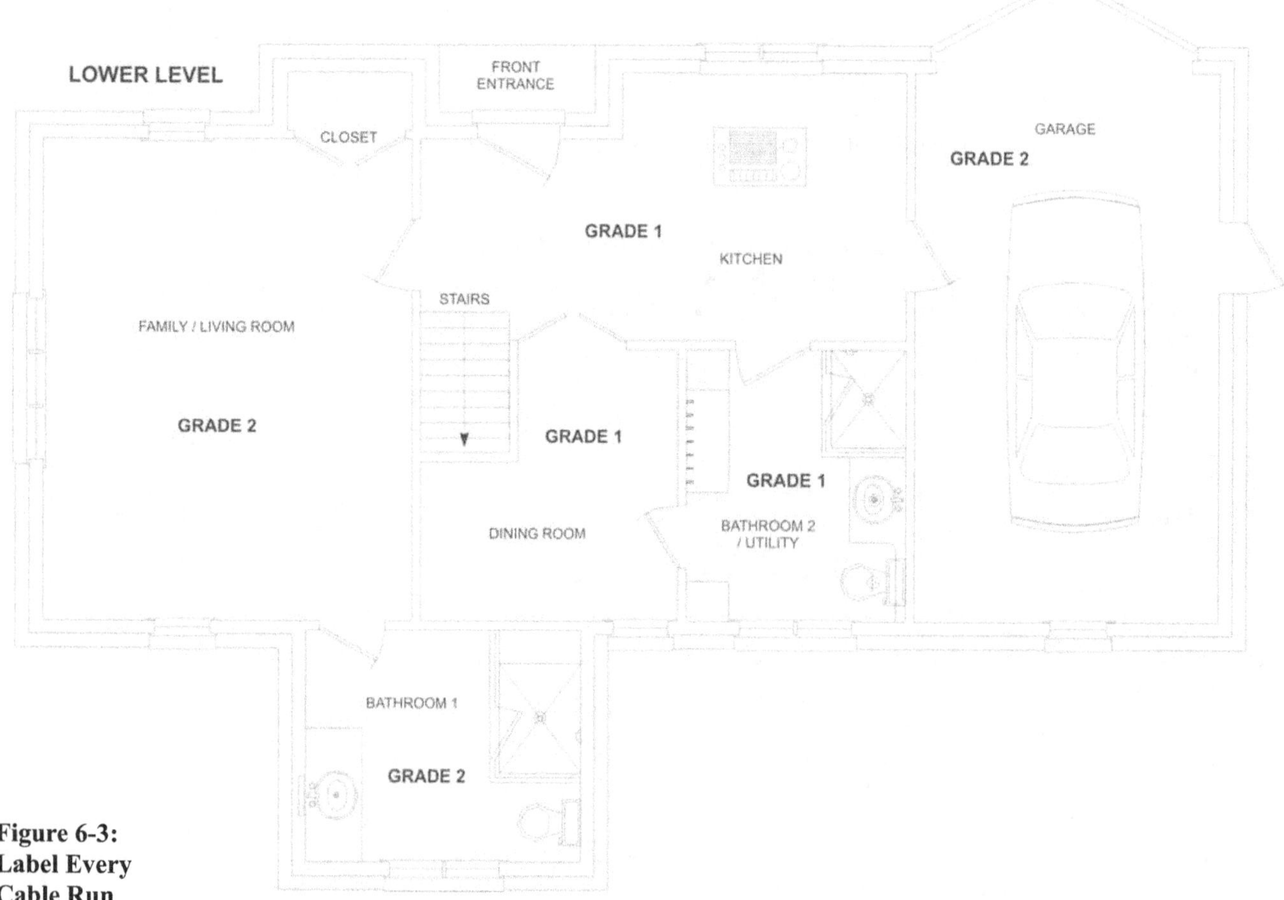

**Figure 6-3:
Label Every
Cable Run**

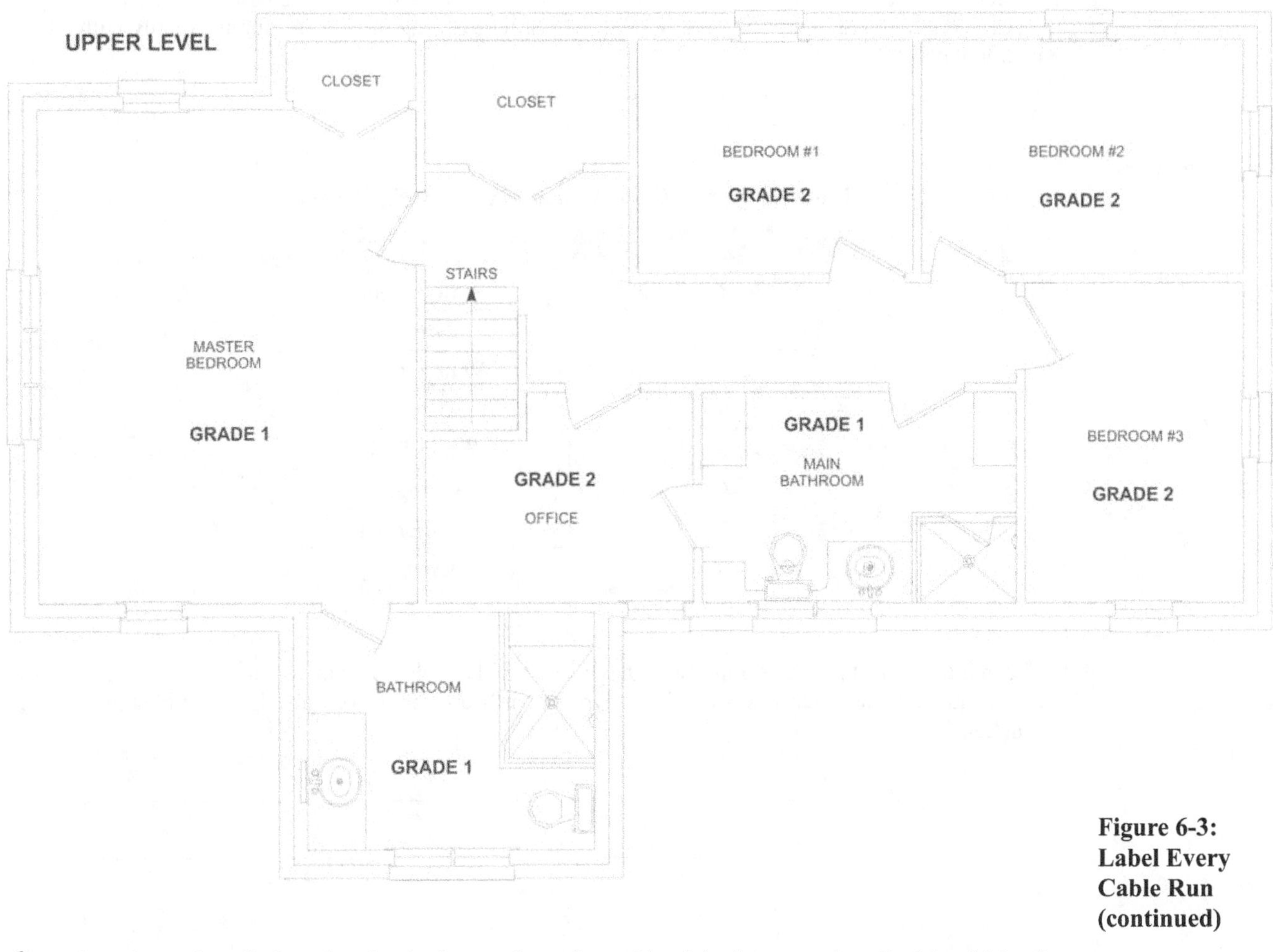

**Figure 6-3:
Label Every
Cable Run
(continued)**

6. The primary installation planning and procedures for residential wiring are described in which of the following standards?

    ___a. TIA/EIA 568A
    ___b. TIA/EIA 569A
    ___c. TIA/EIA 570A
    ___d. TIA/EIA 606

7. Which of the following statements are true about terminating cables with RJ-45 connectors during the installation of a home wiring system?

    ___a. TIA/EIA 568A is the preferred standard for terminating cables with RJ-45 connectors.
    ___b. TIA/EIA 568B is the preferred standard for terminating cables with RJ-45 connectors.
    ___c. TIA/EIA 570A is the preferred standard for terminating cables with RJ-45 connectors.
    ___d. TIA/EIA 606A is the preferred standard for terminating cables with RJ-45 connectors.

8. Proper termination of cables is vital for effective communication across the new system. The color codes and pin assignments outlined in Table 6-6 for terminating the CAT5 cabling comply with which international standard?

____

Table 6-6: CAT5 Color Codes and Pin Assignments

RJ-45 Plug	Wire Color
1	White/Green
2	Green
3	White/Orange
4	Blue
5	White/Blue
6	Orange
7	White/Brown
8	Brown

9. The builder wants to include a hardware device that will enable home buyers to access a whole house audio and video distribution system. What type of hardware should you install to provide this level of functionality?

____

____

Electrical Wiring

As an RESI Environmental Control installer you need know the basics of high-voltage wiring systems.

1. The primary installation planning and procedures for residential electrical wiring systems are described in which of the following standards?

    ____ a. NFPA 1
    ____ b. NFPA 54
    ____ c. NFPA 70
    ____ d. NFPA 101

2. Complete the following statement:

    Adding a home LAN with 1.5 amps of current flow to a 110-volt circuit will add approximately _________ watts to the overall power consumption of the house.

3. You muse ensure that a house's electrical wiring can provide a planned home network with consistent and clean power before installation begins. Identify a simple mechanism for verifying the integrity of the existing wiring system.

____

____

4. The size of a wire determines how much current it can carry to a load. In Table 6-7 rank the current capacity of the various wire types, where 1 is the highest current capacity and 4 is the lowest.

- 14 AWG
- 12 AWG
- 8 AWG
- 10 AWG

Table 6-7: Wire Type Current Capacity

Ranking	Wire Type
1	
2	
3	
4	

5. In Table 6-8 indicate whether the following statements about the installation of electrical cabling are true or false.

Table 6-8: Electrical Cabling Installation Statements

True	False	Statement
		Introducing twists into the cable will affect the performance of the circuit.
		The cable must be handled at the same level of care as CAT5 data cable.
		Introducing sharp turns into the cable will not affect the performance of the circuit.

Access Control Systems

**Power and
Control Systems**

OBJECTIVES

1. Identify a feature of the key control unit that minimizes security risks.
2. Describe the financial impact of using an underground, remote control, hydraulic gate opener.
3. Suggest a low-cost maintenance access control device for each door.
4. Identify components for the monitoring system.
5. Describe the strengths of different types of electromagnetic locks.
6. List organizations that produce standards for home access control systems.
7. Suggest best practices for installing the gate operating system.
8. Identify a termination unit for the low-voltage wires used to connect the various elements of the access control system together.
9. Identify a labeling standard for identifying cables during the installation process.
10. Rank the safety of the various modules covered during the demonstration of the final system.

RESOURCES

1. Marcraft's *RESI Environmental Control Endorsement* textbook
2. Pencil

DISCUSSION

Read the following scenario:

SmartKo is a New York-based company, with over twenty years of experience in the manufacture and installation of automatic gate operators and access control systems. The company has recently won a contract to design and install an access control system for a new office building in the Boston area.

You work remotely for SmartKo as part of its national network of sales experts, and you have been asked by headquarters to advise on product selection and installation of the new system.

Table 7-1 lists general requirements that are included in the contract.

Table 7-1: General Requirements Included in the Contract

Requirement Category	Details
Gate Operator	The gate operating system will need to support swing gates with a width of 10 feet and a weight of 400 pounds. An RF-based key control will be used for opening and controlling the entrance gate.
Entry control	Four exterior perimeter doors require protection. The access control devices mounted on each door will feature low maintenance costs.
Monitoring	Some type of computerized monitoring system is required.
Safety	Full training is required to minimize risks associated with operating the system.

PROCEDURES

Product Selection

Power and Control Systems

1. Your client worries that intruders will use code grabbers to record the code emitted by the control unit when opening the entrance gate. What feature must the key unit support to overcome this potential security breach?

2. What economic impact does selecting an underground, remote control, hydraulic gate opener have on the new access control system?

3. After reviewing the design requirements, can you suggest a low-cost maintenance access control device for each door?

4. Your client wants a monitoring system in order to know exactly what the existing situation is at each door. Which of the following devices will you need to install at each door to provide the monitoring system with status information?

    ____a. Sensors
    ____b. Transformers
    ____c. Actuators
    ____d. Motors

5. Electromagnetic locks will be used to secure the doors. In Table 7-2 match each of the three integrity grades for electromagnetic locks with a description of their locking strength.

- Withstand a pressure of between 500 and 900 pounds.

- Withstand a pressure of between 1,500 and 2,700 pounds.

- Withstand a pressure of between 1,000 and 1,400 pounds.

Table 7-2: Integrity Grades for Electromagnetic Locks

Integrity Grades	Locking Strength
Grade 1	
Grade 2	
Grade 3	

Installation

1. Which of the following organizations produce standards for installing home access control systems?

    ____a. Electronics Industries Alliance (EIA)
    ____b. National Fire Protection Agency (NFPA)
    ____c. Organization for the Advancement of Structured Information Standards (OASIS)
    ____d. Underwriters Laboratories, Inc. (UL)

2. In Table 7-3 indicate whether the following statements about the installation of the gate operating system are true or false.

Table 7-3: Gate Operating System Installation Statements

True	False	Statement
		Be sure that all electrical connections are made in accordance with local electrical codes.
		Use 12 AWG wire for all low-voltage control wiring.
		Secondary entrapment protection devices must be installed to ensure a safe operating environment and to reduce the risk of entrapment.

3. Which type of termination unit is appropriate for terminating the low-voltage wires used to manage the new access control system?

____

____

4. Labeling the wiring for the new access system is one of the final stages of the project. Which of the following statements about labeling and identifying access control cables are true?

    ___a. TIA/EIA 606A is the recommended standard for labeling cables used in access control systems.

    ___b. TIA/EIA 616A is the recommended standard for labeling cables used in access control systems.

    ___c. TIA/EIA 626A is the recommended standard for labeling cables used in access control systems.

    ___d. TIA/EIA 666A is the recommended standard for labeling cables used in access control systems.

5. On completion of the installation and testing, you meet with your client to demonstrate how the new access control system works. In Table 7-4 rank the safety of the different modules of your demonstration, where 1 requires the most attention and 3 requires the least.

Table 7-4: Module Importance

Ranking	Demonstration modules
	Instructions on how to generate reports from the monitoring system.
	Instructions on the required reversing systems associated with the gate operating system and how to test them.
	Instructions on the proper operation of the gate operating system.

Integrating Multizone X-10 Power and Control Systems

OBJECTIVES

1. Integrate the residential power and control system and the OmniLT controller.
2. Implement the operation of the residential power and control system at the Omni console.
3. Test and verify the automated Omni residential power and control system operations.

Power and Control Systems

RESOURCES

1. Marcraft Residential Power and Control Experiment Panel and Frame
2. HAI OmniLT Automation Owner's Manual 21R00-1 Rev. 2.4
3. HAI OmniLT Automation Installation Manual 21I00-1 Rev. 2.4
4. X-10 Power Line Interface Module TW523 or X10 Pro PSC05
5. X-10 interface cable, 6-conductor
6. Computer system with Windows XP Professional, Microsoft Internet Information Service and HAI Web-Link II software installed

DISCUSSION

The OmniLT controller has internal features that permit its integration with X-10 systems. Much of the operational synergy between the OmniLT controller and the X-10 components is similar to that with which you are already familiar, although it is perhaps not as robust. For this procedure, you will not use the computer interface module in conjunction with the Omni system, because the goal here is to run the entire residential operation (all individual Marcraft panels) from one source (the Omni).

NOTE: For the X-10 power line interface module (TW523) to operate correctly in the following Lab. Power to the Network Integration panel must be supplied from the Power and Control Systems panel. Some electrical equipment and computer accessories can absorb the PLC/X10 (Power Line Carrier) signal. The same is also true with other PLC/X10 transmitting devices; up to ½ the signal can be lost to a nearby transmitter.

If at any time the serial cable between computer and the Network Integration Panel becomes dislodged, the computer may have to be restarted to regain control of the serial port. Unlike USB the standard computer serial ports are not HOT swappable.

Because other students will be connecting their respective panels to the OmniLT controller at its panel, this lab procedure will require a high level of teamwork to achieve the desired outcome. To prevent a chaotic situation from developing, your instructor may create a schedule to determine which student team gets access to the Omni panel, and for how long. Marcraft recommends that the residential power and control system not be the first panel to be integrated into the Omni system. It is critical that the residential phone system be connected to the Omni first to prevent the system from locking up or continually reporting errors.

Power and Control Systems

PROCEDURES

Integrating the Residential Power and Control System and the OmniLT Controller

1. Examine the connections to the OmniLT controller panel shown in Figure 8-1.

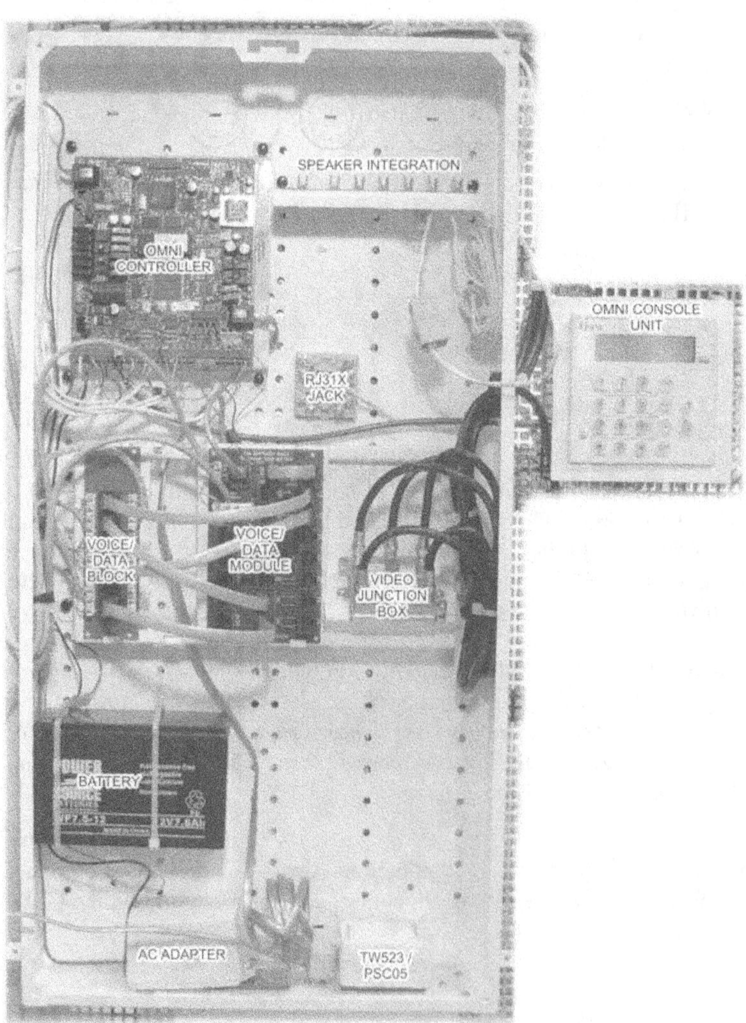

**Figure 8-1:
Connections to the
OmniLT Controller
Panel**

2. Locate the two-way X-10 power line interface module (TW523) and its 6-conductor cable.

The OmniLT controller board should already be powered and actively running at this point.

3. Plug one end of the 6-conductor interface cable into the X-10 jack located on the controller board to the immediate right of the TB3 phone connection block.

4. Plug the remaining end of the 6-conductor interface cable into the jack located at the bottom of the X-10 power line TW523 interface module.

5. Plug the X-10 power line TW523 interface module into an ac power source, preferably at the bottom of the control box as shown in Figure 8-1.

Figure 8-1: Connected TW523 X-10 Module in Control Box

Once the TW523 interface module is plugged in, its LED should glow red.

6. Try to position the module's 6-conductor interface cable so that it does not interfere with other units and/or wiring that may already be deployed in the control box.

7. From the Omni console unit's top-level display, enter the installer setup menu by pressing **9** and providing the appropriate installer code.

8. Press # on the console and view the installer setup menu that appears, similar to Figure 8-2.

Figure 8-2: The Installer Setup Menu

9. Press **1 (CTRL)** and make sure that the specified X-10 house code is *A*. If it is not, adjust it accordingly.

10. Press the **down arrow** key and ensure that the specified X-10 phase setting is **3-phase**. If necessary, adjust this setting to *1* and then press the # key.

Because Marcraft's residential power and control system panel uses a phase coupler, make sure that the X-10 signal will be transmitted along all phases of the electrical system.

11. Press the * key twice to return to the Setup menu.

12. Back at the setup menu, press 6 to get to the miscellaneous section.

13. Press the **down arrow** until you reach the **HC 1 Format:** parameter.

14. Make sure that the setting is **0-STANDARD**.

15. If it is not, press the # key and then press the **0** key.

16. Then, press the # key again to select it.

17. Press the **down arrow**, and at the **HC 1 all Off** parameter, ensure that setting is **1** for *YES*.

18. Press the **down arrow**, and at the **HC 1 all On** parameter, ensure that setting is **1** for *YES*.

These settings allow the X-10 light modules to respond to "All Off" and "All On" commands.

19. Press the * key once to return to the setup menu.

20. Back at the setup menu, press the **7** key to get to the set up name section.

This section enables you to provide descriptive names for the various X-10 units being used by the Marcraft panel. For these units, the names can be up to 12 characters in length.

21. From the *Set Up Name* menu, press the **1** key to begin naming the X-10 modules.

22. When the current name for *UNIT 1* appears, make sure it reads **A1Transcvr**.

23. Use Table 8-1 to determine the numeric codes required to output the name characters.

Table 8-1: Numeric Codes for Character Outputs

Code	Character	Code	Character	Code	Character	Code	Character
00	SPACE	24	8	48	P	72	h
01	!	25	9	49	Q	73	i
02	"	26	:	50	R	74	j
03	#	27	;	51	S	75	k
04	$	28	<	52	T	76	l
05	%	29	=	53	U	77	m
06	&	30	>	54	V	78	n
07	'	31	?	55	W	79	o
08	(	32	@	56	X	80	p
09	)	33	A	57	Y	81	q
10	*	34	B	58	Z	82	r
11	+	35	C	59	[	83	s
12	,	36	D	60		84	t
13	-	37	E	61	$\overline{]}$	85	u
14	.	38	F	62	^	86	v
15	/	39	G	63	_	87	w
16	0	40	H	64	`	88	x
17	1	41	I	65	a	89	y
18	2	42	J	66	b	90	z
19	3	43	K	67	c	91	-
20	4	44	L	68	d	92	x
21	5	45	M	69	e	93	—
22	6	46	N	70	f	94	↕
23	7	47	O	71	g	95	✂

24. If you must edit the name, enter the required two-digit numeric codes for each letter of the name for UNIT 1.

25. Then, press the # key.

26. Use the **down arrow** key and determine the numeric codes required to name *UNIT 2* as **A2Lamp1**.

27. Press the # key and then press the **down arrow** key.

28. Determine the numeric codes required to name *UNIT 3* as **A3KitchenLamp**.

29. Press the # key and then press the **down arrow key** again.

30. Repeat the naming procedure to name *UNIT 4* as **A4Lamp2**.

31. Press the # key again, and then press the **down arrow** key.

32. Name *UNIT 5* as **A5CeilingLt**.

33. Again press the # key followed by the **down arrow** key.

34. Name *UNIT 6* as **A6ShowerFan**.

35. Press the # key followed by the * key.

The OmniLT controller is equipped with recorded voice descriptions that the system can play over the telephone to report status activities. Although the controller's vocabulary is limited, it's large enough to provide meaningful spoken information about the various modules. The language that refers to certain modules mounted on the Marcraft panel may have to be altered slightly in order to utilize the limited vocabulary.

36. From the setup menu, press the **8** key to get to the set up voice section.

37. From the *Set Up Voice* menu, press the **1** key to begin providing voice descriptions for the X-10 modules.

38. When the number for the first unit is displayed, you can scroll through the list of pre-existing voice descriptions using the arrow keys, if any are there.

Remember that UNIT 1 should correspond to the X-10 transceiver module that you've been using to signal when ventilation activity is occurring. However, a "transceiver" voice description is not provided in the list. During the following steps, the Omni console display may time out and return to its top-level menu periodically if no action is taken within a lower level menu. When this happens, you will have to step through the menu system again to get where you were.

39. Examine Table 8-2 to locate the voice description(s) to use for UNIT 1.

Table 8-2: Numeric Codes for Voice Descriptions

Voice Descriptions		Voice Descriptions		Voice Descriptions		Voice Descriptions	
Code	Description	Code	Description	Code	Description	Code	Description
1	ELEVEN	26	A.M.	50	BURGLAR	75	ENERGY
2	TWELVE	27	P.M.	51	BUTTON	76	ENTER
3	THIRTEEN	28	WELCOME TO OMNI	52	BYPASS	77	ENTRY
4	FOURTEEN			53	CANCEL	78	EVENTS
5	FIFTEEN	29	(PAUSE)	54	CENTER	79	EXIT
6	SIXTEEN	30	(SHORT PAUSE)	55	CLOSET	80	FAMILY
7	SEVENTEEN	31	AC POWER	56	CODE	81	FAN
8	EIGHTEEN	32	ACCESS	57	CONTINUE	82	FIRE
9	NINETEEN	33	ADDRESS	58	(BEEP)	83	FOYER
10	TWO	34	ALARM	59	CONTROL	84	FREEZE
11	TWENTY	35	ALL	60	COOL	85	FRONT
12	THREE	36	APPLIANCE	61	DATE	86	FUSE
13	THIRTY	37	AREA	62	DAY	87	GARAGE
14	FOUR	38	ATTIC	63	DEGREES	88	GAS
15	FORTY	39	AUTO	64	DELAYED	89	GIRL'S
16	FIVE	40	AUXILIARY	65	DEN	90	GLASS
17	FIFTY	41	AWAY	66	DENIED	91	GOOD-BYE
18	SIX	42	BACK	67	DIMMER	92	GOTO
19	SIXTY	43	BASEMENT	68	DINING	93	GUEST
20	SEVEN	44	BATH	69	DOOR	94	GUN
21	SEVENTY	45	BATTERY	70	DOWN	95	HAD
22	EIGHT	46	BED	71	DRIVEWAY	96	HALL
23	EIGHTY	47	BOY'S	72	DURESS	97	HEAT
24	NINE	48	BRIGHTER	73	EAST	98	HIGH
25	NINETY	49	BUILDING	74	EMERGENCY	99	HOLD

Table 8-2: Numeric Codes for Voice Descriptions (continued)

Code	Description	Code	Description	Code	Description	Code	Description
100	HOURS	125	NURSERY	149	RECORD	173	TALK
101	HUNDRED	126	OFF	150	REMOTE	174	TAMPER
102	INSTANT	127	OFFICE	151	REPEAT	175	TEMPERATURE
103	INTERIOR	128	OH	152	RESTORE	176	TEN
104	INVALID	129	ON	153	RIGHT	177	THEN
105	IS	130	ONE	154	RISE	178	THERMOSTAT
106	KITCHEN	131	OR	155	ROOM	179	TIME
107	LEFT	132	OUTDOOR	156	SAVER	180	TIMED
108	LEVEL	133	OUTLET	157	SECONDS	181	TO
109	LIGHT	134	PANIC	158	SECURE	182	TROUBLE
110	LISTEN	135	PATIO	159	SECURITY	183	TRIPPED
111	LIVING	136	PC	160	SETTING	184	UNIT
112	LOW	137	PERIMETER	161	SHOP	185	UP
113	MAIN	138	PHONE	162	SIDE	186	VACATION
114	MASTER	139	PLAY	163	SILENT	187	WATER
115	MEDICAL	140	PLEASE CHOOSE	164	SOUTH	188	WEST
116	MINUS	141	POINT	165	SPA	189	WINDOW
117	MINUTES	142	POLICE	166	STAIRS	190	ZONE
118	MODE	143	POOL	167	STAR	191	STOCK
119	MOTION	144	POURCH	168	STATUS	192	UTILITY
120	NIGHT	145	POUND	169	STEPS	193	EQUIPMENT
121	NORTH	146	PRESS	170	STORAGE	194	COMPUTER
122	NOT	147	PUMP	171	SUN	195	APARTMENT
123	NOW	148	READY	172	SYSTEM OK		
124	NUMBER						

40. Pressing # between each code entry, input the **81** and **119** codes from the list of descriptions. Then, press the # key once more to save.

According to the description list, these codes correspond to the words "FAN MOTION", so that when "Unit One" is spoken over the phone line, the OmniLT will say "Unit 1, Fan Motion."

41. Press the **down arrow** to move to the *UNIT 2* entry screen corresponding to the Lamp 1 module that represents the dining room light.

42. Pressing # between each code entry, input the **68**, **155**, and **109** codes from the list of descriptions. Then, press the # key once more to save.

These codes correspond to the voice descriptions "DINING ROOM LIGHT", so that when the words "Unit Two" are spoken over the phone line, the OmniLT will say "Unit 2, Dining Room Light."

43. Press the **down arrow** to move to the *UNIT 3* entry screen and enter the codes **106** and **81**, remembering to press # between each code. Then, press the # key again to save.

44. For the *UNIT 4* description, enter and save the **46**, **155**, and **109** codes.

45. At the menu prompt for UNIT 5's voice description, examine Table 8-2 again.

Unfortunately, the voice descriptions do not include the words "ceiling" or "recreation." However, the word "den" is provided.

46. Enter and save the codes **65** and **109**. Then, move to the next entry.

47. When the menu displays *UNIT 6*, enter and save the codes **44**, **155**, and **81**.

48. Press the * key twice to return to the setup menu. Then, press the * key as necessary to return to the normal top-level display.

The OmniLT controller, console, and Marcraft panel have now been integrated. It's time to try using the console unit to operate the residential power and control system.

Implementing Operation of the Residential Power and Control System at the Omni Console

1. Examine each X-10 module on the Marcraft panel to ensure that all house codes are set to **A**.

2. At the top-level display on the Omni console, press the **4** (*ALL*) key and then press the **1** (*LIGHTS ON*) key. Observe the Marcraft panel.

The console should emit a beep, and all X-10 modules that are lamp types should activate. No X-10 appliance modules or outlets should respond because this command targets lamp modules specifically.

3. On the Omni console, press the **4** (*ALL*) key and then press the **0** (*OFF*) key. Observe the Marcraft panel.

All X-10 modules that are lamp types should deactivate. In addition, any X-10 appliance modules or outlets that were active should now also deactivate. Individual control over the X-10 modules can be examined using the control menu.

4. Press the **1** (*CTRL*) key on the console from the top-level display to enter the control menu.

5. When the first named item in that list is displayed, use the # key to select it. The item text reads:

> A1Transcvr
> 0=OFF 1=ON 9=TIM #=STA ↑

6. If the lamp plugged into the transceiver is not lit, press the **1** (**ON**) key.

The OmniLT console will beep once, and the lamp plugged into transceiver A1 will be turned ON. The console then returns to the top-level display.

7. From the top-level display, return to the control menu by pressing the **1** (**CTRL**) key and select the transceiver module **A1** again.

8. This time press the **0** (*OFF*) key.

The console beeps once, and the lamp plugged into transceiver A1 is turned OFF. The console again returns to the top-level display.

9. Return to the control menu and select the **A2Lamp1 module** by either entering its unit number directly, or by scrolling to it with the arrow keys.

10. Press the # key to select it. The item text should read:

> A2Lamp1
> 0=OFF 1=ON 2=DIM 3=BRT ↓
>
> A2Lamp1
> 4=LVL 5=RMP 9=TIM #=STA ↑

11. If the lamp plugged into the module is not lit, press the ON key **1**.

12. From the control menu, dim the *A2Lamp1* module by pressing the **2** (*DIM*) key. The console display should read:

> A2Lamp1
> STEPS DIMMER (1-9):

13. Press a number between 1 and 9 to indicate the desired dimming.

The console will beep and the light will be dimmed.

14. From the control menu, brighten the A2Lamp1 module by entering its unit number directly, or by scrolling to it with the arrow keys.

15. Then, press the **#** key and press the **3** (*BRT*) key.

16. Press a number between 1 and 9 to indicate the desired brightening, and then press **#**.

17. From the control menu, go to the **A2Lamp1** module and turn it *OFF* by pressing the **0** key.

18. Next, use the control menu to turn the *A3KitchenLamp* **ON** and **OFF** to verify its operation from the OmniLT.

19. Go to the **A4Lamp2** module from the control menu and manipulate it as you did the *A2Lamp1* module to be sure it operates properly.

20. Manipulate the **A5CeilingLt** module from the control menu to ensure it operates correctly.

21. Use the control menu to turn the *A6BathroomFan* **ON** and **OFF** to verify its operation from the OmniLT.

You can specify the length of time that an X-10 module is turned ON or OFF by using timed commands. For timing periods of minutes or seconds, a module may be turned ON or OFF for a count of 1 to 99. When the count has expired, the module will assume its opposite condition. If timing periods of hours are required, values of 1 to 18 are available. X-10 light modules may also be dimmed or brightened for the same specified periods of time. When the specified time period has elapsed, the unit will return to its previous intensity level.

22. From the control menu, enter the **A5CeilingLt** section and press the **#** key.

23. Next, press the **9** (*TIME*) key.

Notices that you can choose between minutes, seconds, and hours.

24. Before entering any digits, use the **#** key to select seconds if necessary.

25. Then, enter a time count of **30**.

Once the time is entered, the control menu is redisplayed with the specified time shown. For example:

```
A5CeilingLt For 30S
0=OFF 1=ON 2=DIM 3=BRT ↓
```

26. Press the **1** key to turn the *A5CeilingLt* ON for 30 seconds.

From the top-level menu, the module should turn OFF after the specified time period has elapsed. These manipulations of the OmniLT controller, console, and the Marcraft panel have shown that the power and control system is now operational. You will now create a useful program to automatically operate it.

Testing and Verifying Automated Omni Residential Power and Control System Operations

Automated control functions can be programmed into the Omni system according to a time schedule, or in response to a system event. For example, a program can be set to automatically execute:

- once at a certain time on a certain date
- on a certain date every year
- repeatedly
- in response to an event
- only under certain circumstances

You will program the OmniLT controller to automatically execute a program, similar to what was accomplished when the Marcraft power and control panel operated as a stand-alone unit. Because of differences in the OmniLT controller's design, identical operations will not be possible. The main difference is the way in which the designers of the Omni system interpret the word "program." Their understanding is that a single step constitutes a program by itself, because it contains information about when the specified action should occur, what the specified action is, and what specific condition must exist in order for the expected action to occur.

The following steps are intended to mimic the program you created earlier, although it will not be able to copy it exactly. You may want to have Table 8-3 handy as you perform them.

1. From the top-level menu, press the **9** (*SETUP*) key on the console and enter the appropriate installer code.

For adding, reviewing, changing, and deleting automation programs, you need to be at the setup program menu.

2. From the setup menu, press the **3** (*PROG*) key to get to the program menu. The console display should read:

        SET UP PROGRAMS
        1=ADD 2=SHOW 3=DELETE

3. To add a new automation program to the system, press the **1** (*ADD*) key. The edit program menu appears as:

        EDIT PROGRAM
        1=WHEN 2=CMD 3=&COND

4. At the edit program menu, press the **1** (*WHEN*) key.

5. At the edit when menu, press the **1** (*TIMED*) key.

6. Then, ignore any default data and press the **1** (*TIME*) key to enter the time you want the first program command to execute.

This entry will depend on what hour your class meets. For a morning class you'll want to select the time between 8:00 AM and 11:00 AM during which your class is in session. For an afternoon session, select a time between 1:00 PM and 4:00 PM. For purposes of this procedure, we'll assume an early morning class is in session and set this time to 9:00 AM. This means that the first program step would execute at 9:00 AM. Make sure you leave enough time to test the program by specifying times at which you will be present when it executes.

7. Enter the time **9:00** and press the **up arrow** for **AM** (RISE/AM).

8. Back at the timed menu press the **2** (*DATE/DAY*) key.

This enable you to select a specific date or certain days of the week for program execution. The console displays:

        DATE: 11/17
        MMDD ↓=DAY

9. Press the **down arrow key** on the console.

10. Then, press **8** to apply the input time to the weekdays (Monday through Friday).

11. Press the # key twice to return to the edit program menu.

12. From the edit program menu, press **2** (*CMD*) to specify the first command to be performed when the program executes at 9:00 AM. The console displays:

        1=CONTROL 2=SECURITY
        3=BUTTON 4=ALL   ↓

        5=TEMP 6=ENERGY
        8=MESSAGE   ↑

13. Because you are applying commands to a lamp module, press the **1** (*CONTROL*) key.

14. Locate the **A4Lamp2** module in the listing. Press the # key and the console should appear as:

        A4Lamp2
        0=OFF 1=ON 2=DIM 3=BRT ↓

        A4Lamp2
        4=LVL 5=RMP 9=TIM #=STA ↑

15. Press the **1** (**ON**) key to obtain a brightness level of **100%**.

16. Back at the edit menu, press the **3** (*&COND*) key to specify what pre-existing conditions to look for at the *A4Lamp2* module prior to program execution.

17. Press the **1** (*CTRL*) key and enter the module's unit number (**4**). Then, press the # key.

18. When the console prompts you to enter the prior state of the *A4Lamp2* module, select **0** (*OFF*).

19. Back at the edit program menu, press the # key to display the program step you just created, and press the # key again to save it.

A successful save of the step, or "program," will return the display to the set up programs menu. From here, you go to the next program, or step. The second step in your previous program was turning the shower fan/transceiver combination ON.

20. From the set up programs menu, press the **1** (*ADD*) key to begin programming the next step.

21. When the edit program menu appears, press the **1** (*WHEN*) key.

22. At the edit when menu, press the **1** (*TIMED*) key, and recall the previous program command execution time.

23. Press the **1** (*TIME*) key to enter the time you want the second program command to execute.

This entry will depend on the start time you selected for the first program step. For purposes of this procedure, we chose 9:00 AM. Following the format established for the program you created earlier, the second program step would execute at 9:01 AM.

24. Enter the time **9:01** and press the **up arrow** key for **AM** (RISE/AM). Then, from the timed menu press the **2** (*DATE/DAY*) key.

25. Press the **down arrow** key on the console followed by **8** to apply the input time to Monday through Friday.

26. Return to the edit program menu by pressing the # key twice.

27. Press **2** (*CMD*) to specify the command to be executed at 9:01 AM. Then, press the **1** (*CONTROL*) key.

28. Locate the A6ShowerFan module in the listing. Press the # key console should appear as below:

```
        A6ShowerFan
        0=OFF 1 = ON 2=DIM 3=BRT   ↓

        A6ShowerFan
        4=LVL 5=RMP 9=TIM #=STA   ↑
```

29. Press the **1** (ON) key

30. From the edit menu press the **3** (&COND) key, then press **1** (CTRL) key, enter the unit number (**6**), and press the # key.

31. At the prompt, select **0** (*OFF*) as the prior state of the *A6ShowerFan* module.

32. From the edit program menu, press the # key to see the program step and press # again to save it.

Remember that the stand-alone version of this program activated the transceiver light whenever fan operation was required.

33. From the set up programs menu, press the **1** (*ADD*) key. At the edit program menu press the **1** (*WHEN*) key.

34. At the edit when menu, press the **1** (*TIMED*) key and recall the previous program command execution time.

35. Because the transceiver light parallels fan operation time, press the **1** (TIME) key and enter **9:01** again, Then press the **up arrow** key for **AM** (RISE/AM).

36. From the timed menu press the **2** (*DATE/DAY*) key, the **down arrow** key, and **8** for *Monday through Friday* operation.

37. Press the # key twice to return to the edit program menu.

38. Press **2** (*CMD*) and specify that this command will also execute at *9:01 AM*. Then, press the **1** (*CONTROL*) key.

39. Locate the *A1Transcvr* module in the listing. Press # key console should appear as below: then and press **1** (ON) key.

```
A1Transcvr
0=OFF 1 = ON 2-5=A-D
```

40. From the edit menu press the **3** (*&COND*) key, enter the unit number (**1**), and press the # key.

41. At the prompt, select **0** (*OFF*) as the prior state of the *A1Transcvr* module.

42. From the edit program menu, press the # key to see the program step and press # again to save it.

As you can already see, getting the OmniLT to mimic the program created earlier will require considerable patience and effort. Complete as many steps as your class time will allow. Manually check each step with a pencil mark to keep track of where you are. Do not be surprised to discover that insufficient class time exists to complete the programming steps.

43. Program the OmniLT to turn **OFF** the *A6ShowerFan* and *A1Transcvr* light at time **9:02AM**.

44. At program time 9:03 AM direct Omni to **DIM** the *A4Lamp2* module to **60%** brightness.

45. At program time 9:04 AM direct Omni to **DIM** the *A4Lamp2* module to **0%** brightness.

46. At program time 9:05 AM direct Omni to turn the *A3Kitchen Lamp* module and the *A1Transcvr light* **ON**.

Be aware that when the phase coupler is operating, the split X-10 signal will cancel itself out as it passes through the coupler. This is because the phases supplied to the X10 Pro 240 Vac X-10 outlet are placed 180 degrees out of phase by the power transformer. Deactivating the phase coupler at the main panel will alleviate this problem.

47. At program time 9:06 AM direct Omni to operate the *A2Lamp1* module at **60%** brightness.

48. At program time 9:07 AM direct Omni to turn the *A3KitchenLamp* module and the *A1Transcvr light* **OFF**.

49. At program time 9:08 AM direct Omni to **DIM** the *A2Lamp1* module to **0%** brightness.

50. At program time 9:09 AM direct Omni to operate the *A5CeilingLt* at **100%** brightness.

51. At program time 9:10 AM direct Omni to **DIM** the *A5CeilingLt* module to **0%** brightness.

52. At program time 9:11 AM direct Omni to turn the A6ShowerFan module and the A1Transcvr light ON.

53. At program time 9:12 AM direct Omni to turn the A6ShowerFan module and the A1Transcvr light OFF.

54. At program time 9:13 AM direct Omni to operate the *A5CeilingLt* at **60%** brightness.

55. At program time 9:14 AM direct Omni to **DIM** the *A5CeilingLt* module to **0%** brightness.

56. At program time 9:15 AM direct Omni to turn the *A3KitchenLamp* module and the *A1Transcvr light* **ON**.

57. At program time 9:16 AM direct Omni to turn the *A6ShowerFan* module **ON**.

58. At program time 9:17 AM direct Omni to turn the *A6ShowerFan* module **OFF**, and to operate the *A2Lamp1* module at **100%** brightness.

59. At program time 9:18 AM direct Omni to turn the *A3KitchenLamp* module and the *A1Transcvr light* **OFF**.

60. At program time 9:19 AM direct Omni to **DIM** the *A2Lamp1* module to **70%** brightness.

61. At program time 9:20 AM direct Omni to turn the **A3KitchenLamp** module and the *A1Transcvr light* **ON**. Also **DIM** the *A2Lamp1* module to 60% brightness.

62. At program time 9:21 AM direct Omni to operate the *A5CeilingLt* at **100%** brightness.

63. At program time 9:22 AM direct Omni to turn the *A6ShowerFan* module **ON**.

64. At program time 9:23 AM direct Omni to turn the *A3KitchenLamp* module **OFF**.

65. At program time 9:24 AM direct Omni to **DIM** the *A2Lamp1* module to **0%** brightness.

66. At program time 9:25 AM direct Omni to turn the *A6ShowerFan* module and the *A1Transcvr light* **OFF**.

67. At program time 9:26 AM direct Omni to **DIM** the *A5CeilingLt* to **60%** brightness.

68. At program time 9:27 AM direct Omni to operate the *A4Lamp2* module at **100%** brightness.

69. At program time 9:28 AM direct Omni to **DIM** the *A5CeilingLt* to **0%** brightness.

70. At program time 9:29 AM direct Omni to **DIM** the *A4Lamp2* module to **60%** brightness.

71. At program time 9:30 AM direct Omni to **DIM** the *A4Lamp2* module to **0%** brightness.

72. Go to the set up programs menu and press the SHOW key **2**.

Remember that this menu is where you can review, edit, or delete existing programs in the OmniLT. At this point you should review the program(s) you just entered.

73. From the show programs menu, press the **1** (*CTRL*) key.

74. Use the arrow keys to scroll through the various program commands and ensure that they are correct and complete.

75. To make a necessary correction to a program command, press the # key at its display. The console display reads:

```
SHOW PROGRAM
1=EDIT 2=DELETE
```

76. Press the **1** (*EDIT*) key and make the necessary correction(s). Then, press # to review and # again to save the correction.

77. At the appointed start time of 9:00 AM (or your specified time) observe the Marcraft panel for evidence of program operation.

78. As your program executes, make sure that the instructor knows that the Marcraft power and control panel is working properly with the OmniLT controller.

Modern X-10 systems enable clients to operate the system from a remote location through a touch-tone phone. The following steps will be performed using a touch-tone phone, not the integration controller keypad!

79. Ensure that the integration controller's telco line (white/blue, blue) is connected to the **CO LINE 8** phone line simulator's **right** port.

80. Next, connect the **Remote Location 1** phone to the **CO LINE 8** phone line simulator's **left** port.

81. Examine Figure 8-3 to check your connections.

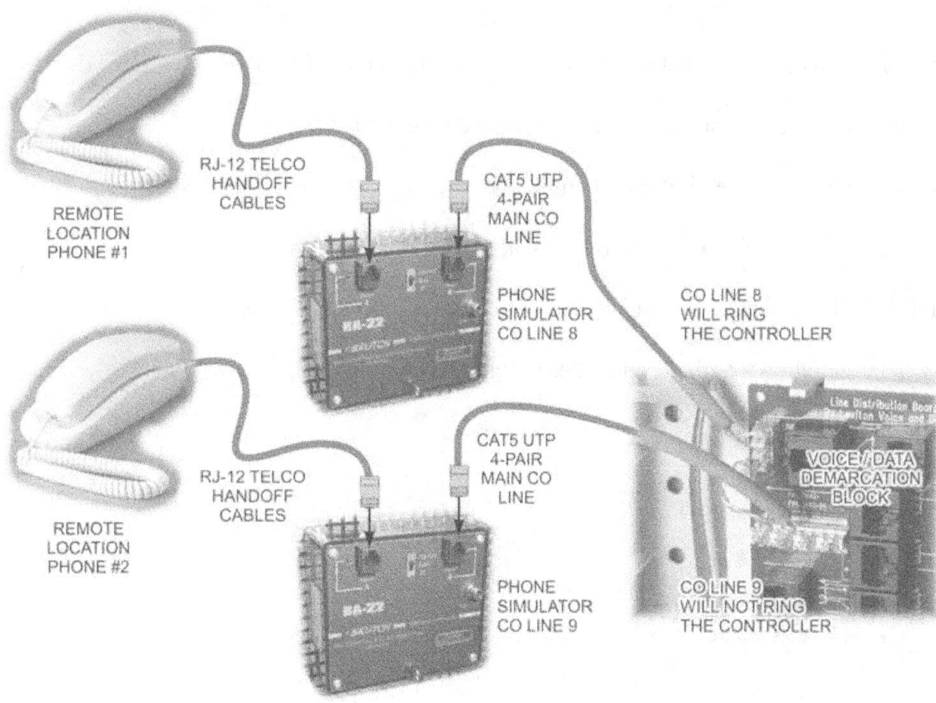

Figure 8-3: Touch-Tone Phone and Controller Connection Setup

The OmniLT controller's main menu will be contacted through the touch-tone phone.

82. At the phone line simulator panel, pick up the **Remote Location 1** telephone handset.

83. Allow the phone to provide the ring signal until the OmniLT controller beeps.

84. Within the next 3 seconds, enter the **master code** supplied by your instructor.

85. Listen for the voice message "Welcome to Omni. Please choose…" to be spoken in the handset.

86. Now, listen to the following menu selections provided by Omni:

1 CONTROL
2 SECURITY
3 BUTTON
4 ALL
5 TEMPERATURE
6 STATUS
7 EVENTS
8 PHONE
9 GOOD-BYE
* CANCEL
0 REPEAT

87. Enter into the All submenu by pressing the **4** (*ALL*) phone button.

88. Listen for the voice message "Please choose..." to be spoken in the handset.

89. Now, listen to the following menu selections provided by Omni:

0 ALL OFF
1 ALL ON
2 LIGHT SETTING

90. Select option **0** and ensure all that all of the X-10 units are in the **OFF** mode.

91. Back at the previous menu, press the **1** (*CONTROL*) phone button.

92. Listen for the voice message "Enter unit number and #" to be spoken in the handset.

93. Now, enter **1#** on the phone keypad to refer to the *A1 Transcvr* unit.

94. Listen for the voice message "Please choose..." to be spoken in the handset.

95. Now, listen to the following menu selections provided by Omni:

0 OFF
1 ON
2 DIMMER
3 BRIGHTER
4 LEVEL
9 TIME
STATUS
* CANCEL

Remember that some of the listed options may not apply to the type of device being manipulated because certain X-10 units have unique operating parameters.

96. Press the **9** (*TIME*) phone button.

97. Listen for the voice message "Enter minutes then #, * Cancel…" to be spoken in the handset.

98. Next, enter **1#** on the phone keypad to select the **A1Transcvr** unit for **one** minute of operation.

99. After the voice message "Time for one minute; Please choose..." is spoken, listen to the following menu selections:

```
0  OFF
1  ON
2  DIMMER
3  BRIGHTER
#  STATUS
*  CANCEL
```

100. Enter **1** on the phone keypad to turn the *A1Transcvr* unit **ON** for **1** minute.

When the voice message, "ON for one minute" is heard, the *A1Transcvr* unit is turned ON.

101. When the Omni reverts back to its main menu, wait for the *A1Trancvr* unit to turn OFF.

102. Now, use the phone keypad to turn the *A2Lamp1 unit 2* **ON** for one minute.

103. Observe the *A2Lamp1* unit and wait for it to turn OFF.

104. Once the *A2Lamp1* unit turns OFF, use the phone keypad to turn the *A3KitchenLamp unit 3* **ON** for one minute.

105. Wait for the *A3KitchenLamp* to shut OFF.

106. Once the *A3KitchenLamp* unit turns OFF, use the phone keypad to turn the *A4Lamp2 unit 4* **ON** for one minute.

107. Wait for the *A4Lamp2* to turn OFF.

108. Then, turn the *A5CeilingLt unit 5* **ON** for one minute using the phone keypad.

109. Wait for the *A5CeilingLt* to turn OFF.

110. Using the phone keypad, turn the *A6ShowerFan unit 6* **ON** for one minute, then wait for it to turn OFF.

111. Once the instructor has observed proper program operation and the program executes the final program step, hang up the Remote Location 1 telephone.

When a system such as the OmniLT is coupled with software that permits its remote monitoring and operation over the Internet, the client can always communicate with his or her residential power and control devices from any location in the world.

While remote monitoring and operation can be performed from any classroom computer network with Web access, the current configuration enables you to perform the following steps directly from the computer running the OmniLT controller.

112. Make sure that the computer system and the X-10 panel are still running. If they are not, power them up.

113. Verify that the wire connections between the Marcraft X-10 panel and the OmniLT controller panel are still intact.

114. Use the Omni console's keypad to send an **All Off** signal to the X-10 devices on the Marcraft panel.

115. On the computer running the OmniLT controller, start the **Web-Link Server** program if necessary.

If the Web-Link Server program is already running, its house icon will appear in the taskbar.

116. On the desktop, click the **Web-Link web client** icon.

117. In the *HAI Web-Link II* login screen, enter the **User Name** and the **User Code**, as shown in Figure 8-4.

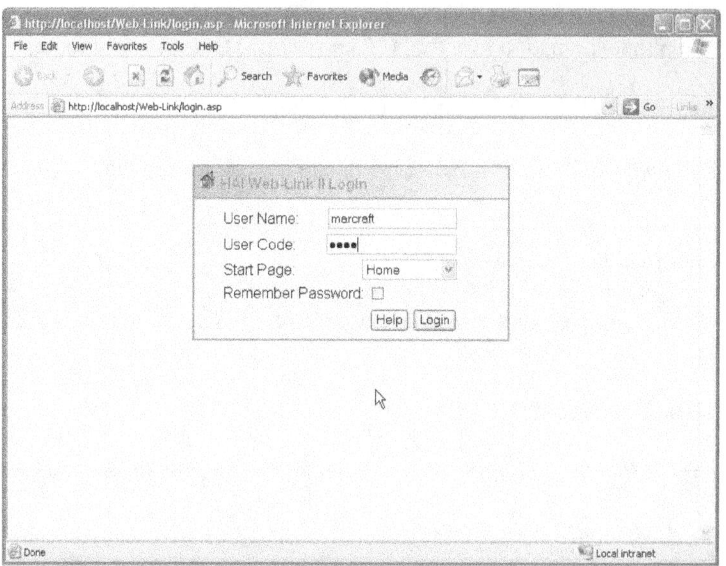

**Figure 8-4:
HAI Web-Link II
Login Screen**

You must enter this information correctly because only three attempts are allowed. If these attempts are erroneous, the system will issue a one-hour lockout.

118. Click on the **Login** button.

119. Then, click the **Control** button to get to the window shown in Figure 8-5.

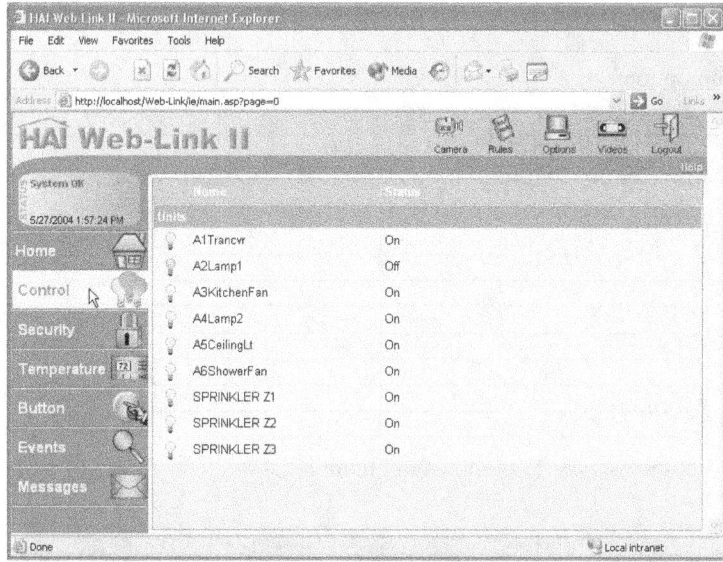

Figure 8-5:
Control Button
Window

120. In the right window pane, click the word **A1Trancvr**.

121. Click on the **down arrow** button next to the word **Command** to display the available options, as shown in Figure 8-6.

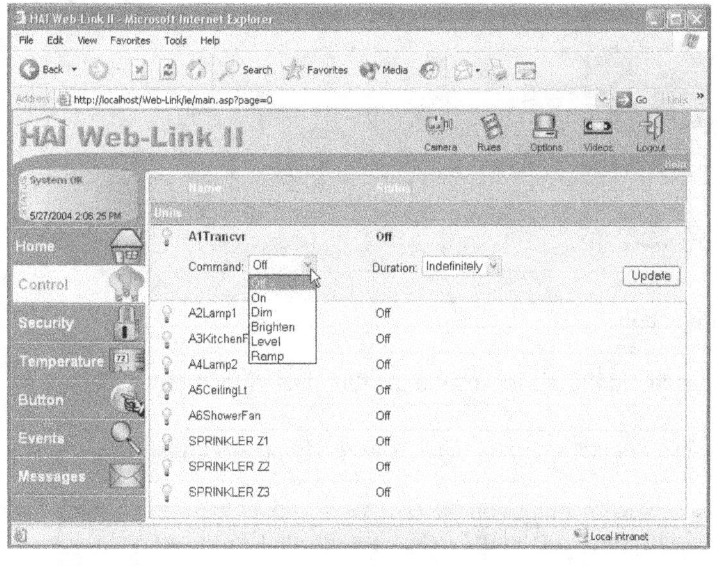

Figure 8-6:
Command Options
for A1Trancvr

122. Record the available command options in Table 8-3.

Table 8-3: Available
Command Options

123. Select the **ON** command option.

124. Select the **Duration** option.

125. Record the duration options in Table 8-4.

Table 8-4: Available Duration Options

126. Set the duration to **Minutes**.

127. For the number of minutes, type **1**, as shown in Figure 8-7.

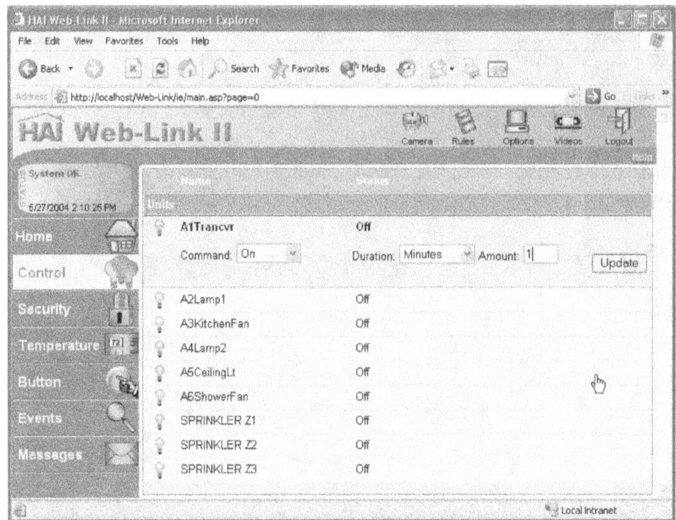

Figure 8-7: Changing Duration for A1Trancvr

128. Click on the **Update** button.

At this point, the A1Trancvr should illuminate and remain illuminated for 1 minute.

129. Using the previous series of steps turn on the **A2Lamp1** for **10 seconds**.

130. Continue to conduct remote operations on the HAI Web-Link by repeating the previous series of steps for the **A3KitchenLamp**, **A4Lamp2**, **A5CeilingLt**, and **A6ShowerFan**.

131. When you have ascertained that all of the X-10 modules can be controlled remotely with the HAI Web-Link software program, click the **Logout** button at the top right of the browser window, as shown in Figure 8-8.

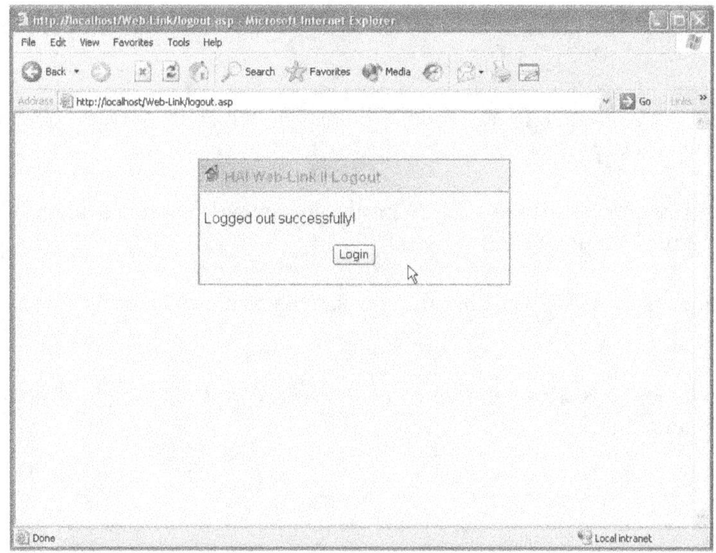

Figure 8-8:
Logged Out
Successfully

132. Use the Omni console to enter the set up programs menu, and press the **3** (*DELETE*) key. The console displays:

> DELETE ALL PROGRAMS?
> 0=NO 1=YES

133. To permanently delete all of the automation programs in the system, press the **1** (*YES*) key.

134. Turn the Marcraft power and control system panel **OFF** by deactivating its red power indicator.

135. Perform a general cleanup of your lab area, and return all applicable equipment and components to their proper storage locations.

LAB QUESTIONS

1. How many conductors does the cable linking the two-way X-10 power line interface module (TW523) to the OmniLT controller board have?

2. What value was specified for the X-10 phase setting?

3. What settings allow the X-10 light modules to respond to "All Off" and "All On" commands?

4. How many characters in length can the descriptive names for the various X-10 units be?

5. What was the descriptive name given to the Omni for X-10 module "UNIT 3:"?

6. What voice descriptions were selected for the "UNIT 2:" Lamp 1 module?

7. What voice descriptions were selected for the "UNIT 5:" Ceiling Light module?

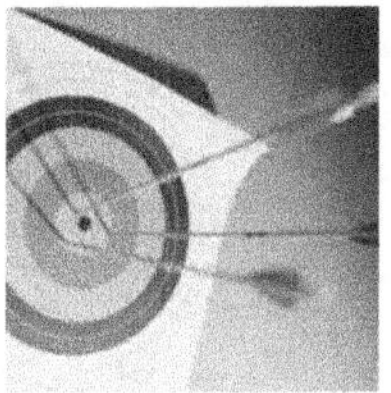

Feedback

8. Why don't X-10 appliance modules or outlets respond to the "All On" command?

9. How many dimming levels does the OmniLT provide for X-10 lamp modules?

10. Which menu is necessary for adding, reviewing, changing, and deleting automation programs in the OmniLT?

11. When creating or editing a program in the OmniLT controller, which three types of information are required for each program step?

12. How can you write programs for the OmniLT controller to achieve the same amount of activity, and yet do it in about half the programming steps we used?

13. How many attempts does the HAI Web-Link II login screen permit for entering the User Name and the User Code?

14. Which of the following options are not permitted by the HAI Web-Link program: monitoring, controlling, or programming?

HVAC System Connections

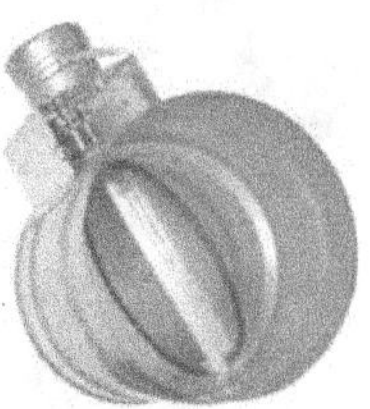

HVAC Systems

OBJECTIVES

1. Identify and describe the function of main components included in HVAC systems.
2. Identify wire connection points between the HVAC controller and the furnace/air handler.
3. Recognize the wire color-coding scheme for the HVAC controller and furnace/air handler.
4. Interconnect the parts discussed within this lab procedure.
5. Identify and describe the function of an HVAC system thermostat.
6. Identify wire connection points between the HVAC controller and the thermostat.
7. Recognize the wire color-coding scheme for the HVAC controller and the thermostat.
8. Identify and describe the function of an HVAC system sensor.
9. Identify wire connection points between the HVAC controller and the sensors.
10. Recognize the wire color-coding scheme for the HVAC controller and the sensors.
11. Identify and describe the function of an HVAC system power source.
12. Identify wire connection points between the HVAC controller and power source.
13. Recognize the wire color-coding scheme for the HVAC controller and the power source.

RESOURCES

1. Marcraft HVAC Experiment Panel and Frame

TOOLS

1. Flat-tip screwdriver, precision head
2. Philips screwdriver, standard 6 inch

DISCUSSION

HVAC controllers come in various configurations and zoning abilities. It is important to consider the needs of the customer as well as the minimal heating and cooling requirements of the building. Multilevel structures or sprawling ranch-style buildings usually require that a system with zoning capabilities be used over a non-zoned system. If a chosen system is insufficient for the building, then temperature differences will occur in various parts of the building. This temperature variance can cause undue wear on the devices and shorten their life expectancy.

A communication connection between the controller and the heating or cooling device should require very little wiring.

Standard AWG 18, 5-conductor wire will be sufficient to connect between the controller and each device. The communication wires are connected to a contact switch. This switch, when activated by the controller, creates a contact path to supply the heating or cooling device with the necessary operational voltage.

PROCEDURES

Furnace/Air-Handler Connections to the HVAC Controller

NOTE: Pin 5 of the controller's HVAC System will not be used to connect a compressor to the controller. Since we do not use a compressor in our environmental control model, we will not use this attachment point. The cooling cycle calls for the air handler fan to operate with or without the activation of a compressor.

The Marcraft furnace/air handler is wired internally to prevent the heating element from operating without the activation of the air handler fan. Connection of the furnace signal wires to the controller will force the operation of the air handler. The air handler is also wired to operate independently. Connection of the fan signal cables will only induce operation of the air handler.

1. Review the manufacturer's installation and operation manual for the HVAC controller along with the connection overview in Figure 9-1.

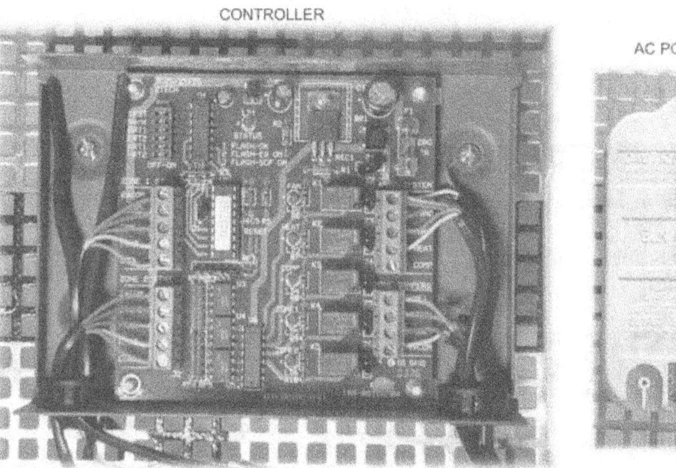

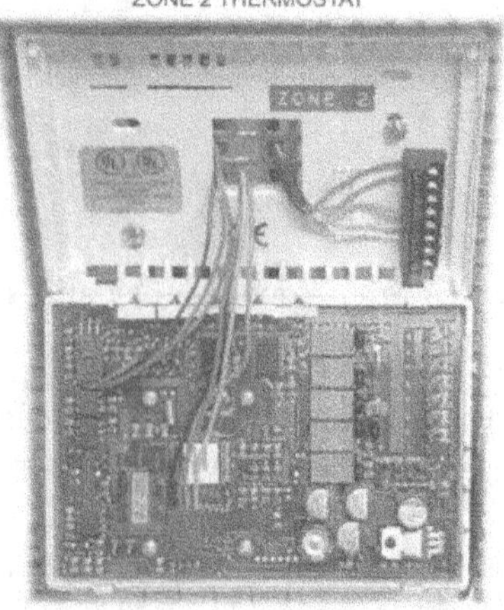

Figure 9-1: Component Connection Overview

2. Use the information supplied by the manufacturer, and Figures 9-1 and 9-2, to complete the appropriate entries in Table 9-1.

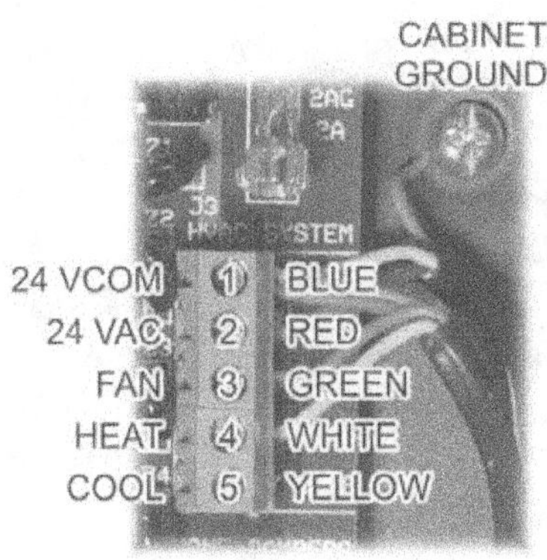

**Figure 9-2:
Connection Diagram
for HVAC System**

Table 9-1: Connection Color Chart for HVAC System

HVAC System		Wire Color Air-Handler	Wire Color Furnace	Wire Color AC Power
PIN 1	24VCOM			
PIN 2	R24VAC			
PIN 3	G FAN			
PIN 4	W HEAT			
PIN 5	Y COMP	NOT USED	NOT USED	
Cabinet Ground				

3. Have your instructor verify the contents of Table 9-1 for accuracy.

4. Make the wire connections to the HVAC controller's HVAC SYSTEM terminal block.

5. Carefully check your wiring against the example in Figure 9-3.

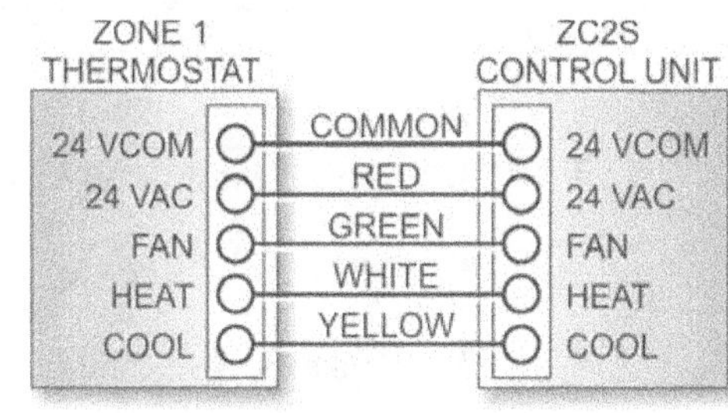

**Figure 9-3:
Wiring Diagram**

Thermostat Connections to the HVAC Controller

Connections between the thermostat and the controller require only two wires per device. These connections require a signal wire and a shared common ground. The thermostat signals the controller to activate the connection being called. It is normal to use 18/5 AWG power-limited circuit cable for a thermostat to the HVAC controller. Five wires are needed because more than one device can be called at a time. Color-coding for this type of wire is fairly standard. The color coding requirements are visible on the printed circuit boards for both devices.

1. Review the installation and operation manuals for the HVAC controller and thermostat.

2. Use the information supplied by the manufacturer, along with information from Figures 9-1 and 9-4, to complete Table 9-2.

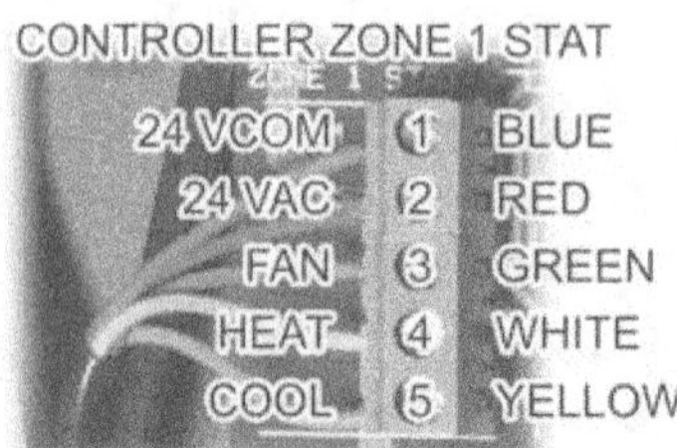

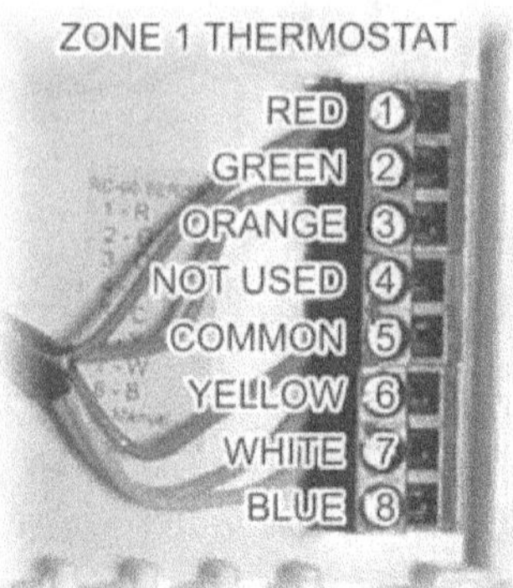

Figure 9-4: Connection Diagram for ZONE 1 STAT to Zone 1 Thermostat

Table 9-2: Connection Color Chart for ZONE 1 STAT to ZONE 1 Thermostat

ZONE 1 STAT		Wire Color from Thermostat	Thermostat 1		Wire Color from Controller
PIN 1	24COM B		PIN 1	1-R	
PIN 2	24VAC R		PIN 2	2-G	
PIN 3	FAN G		PIN 3	3-O	
PIN 4	HEAT W		PIN 4	4-n/u	
PIN 5	COOL Y		PIN 5	5-C	
			PIN 6	6-Y	
			PIN 7	7-W	
			PIN 8	8-B	

3. Have the instructor verify the contents of Table 9-2 for accuracy.

4. Make the wire connections to the HVAC controller's ZONE 1 STAT terminal block.

5. Make the wire connections to the Zone 1 thermostat's terminal block.

Remote Sensor Connections to the Thermostat

Remote sensors are often used with controlling devices such as thermostats. It is through the actions of sensors that controlling thermostats decide when and how to activate the HVAC system.

It is important to use the proper signal cable when installing remote sensors. 18/2 AWG shielded thermostat cable should be used for making these connections to the thermostat. The shielding is vital in protecting the sensor signals to the thermostat. These cables must also run parallel to electrical cables and cross electrical cables perpendicularly to prevent electromagnetic interference.

The wire colors for the type of cable being used may differ depending on the manufacturer. The values for Table 9-3 will depend upon these conditions.

1. Review the manufacturers' installation and operation manuals for the thermostat and remote temperature sensor.

2. Use the information supplied by the manufacturer and Figure 9-5, to complete Table 9-3.

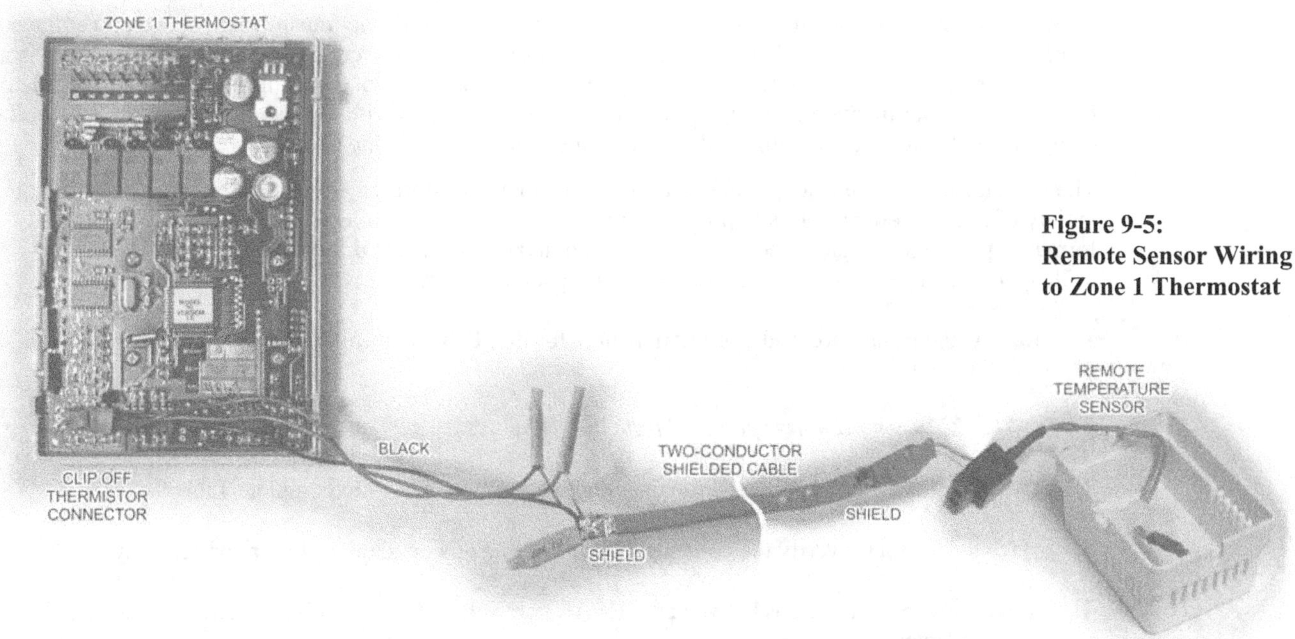

Figure 9-5:
Remote Sensor Wiring
to Zone 1 Thermostat

Table 9-3: Connection Block Diagram for Remote Sensor to ZONE 2 Thermostat

	Wire Color from Remote Sensor Connector	Wire Color from Thermostat's J4 Connector
PIN 1		
PIN 2		
Shield		

3. Have the instructor verify the contents of Table 9-3 for accuracy.

4. Use wire nuts to make the connections between the 2-wire shielded cable and the remote sensor connector.

5. Use wire nuts to make the connections between the 2-wire shielded cable and the Thermostat's J4 connector.

6. Attach the J4 connector to the thermostat.

7. Mate the thermostat to its housing and clip it into place.

AC Power Adapter Wiring to the Controller

HVAC controller manufacturers provide specific requirements for the ac transformer used to power the device. The voltage and amperage draw of the system must be taken into consideration when the system is in the design phase. If the transformer cannot supply the necessary current draw for the system devices then the system will not function properly. It is best to design the system with a standard voltage need. The other components should be compatible with the controller whenever possible.

The manufacturer may not supply the controller with an ac transformer. However, they are available from a variety of different sources. It is important to read the manufacturer's literature pertaining to the ac transformer requirements. Understand the minimum and maximum current draws for the devices being controlled and use these requirements to obtain the proper ac transformer.

1. Review the installation and operation manual for the HVAC controller, the power adapter literature, and Figure 9-3.

NOTE: The ZC2S control unit is powered from the HVAC system's 24Vac and 24VCOM terminals.

2. Use the information supplied by the manufacturer and Figure 9-3 to complete Table 9-1.

3. Have the instructor verify the contents of Table 9-1 ac power adapter section for accuracy.

4. Make the wire connections between the HVAC controller's HVAC SYSTEM terminal block and the power adapter.

5. Ensure that no power is being supplied to the duplex commercial grade electrical receptacle via the single-pole general-use ac switch.

6. Plug the 24 Vac transformer into the duplex commercial grade electrical receptacle.

Feedback

LAB QUESTIONS

1. Why is it necessary to use shielded cable when making connections between a thermostat and a remote sensor?

2. What building structure(s) would indicate that a multizone HVAC system should be used?

3. What is the purpose of the contact switch used in the furnace/air handler?

4. Why is it important to get the correct ac transformer for the HVAC controller?

5. What is the standard wire type used between the thermostat and the controller?

Testing a Single-Zone System

OBJECTIVES

1. Identify and describe the function of a single-zone HVAC system.
2. Test the wire connections to the controller from the furnace/air-handler, thermostat, and remote sensor.
3. Test the thermostat's ability to control the heating and cooling needs of a single room.

HVAC Systems

RESOURCES

1. Marcraft HVAC Experiment Panel and Frame

TOOLS

1. Flat-tip screwdriver, precision head
2. Philips screwdriver, standard 6 inch

DISCUSSION

Control of centralized heating and cooling systems is generally provided at two levels as follows:

- *Overall control* — Overall control is provided by centralized thermostats that measure the temperature at key points within the residence, and turn the heat/cooling on or off at some predetermined set point.

- *Localized control* — Localized control is provided by adjustable airflow vents installed in the registers (exit vents of the ductwork). These devices control the volume of warm or cool air that can enter the room over time. Localized control fine tunes the temperature control function by matching air volume against the natural heat-up and cool-down characteristics of the room.

The following procedure will show just how limited the control is over a single-zone system. Only a single thermostat is used to service the entire structure. The temperature sensor may not adequately cover the structure. If a single-zone system is installed in a multiroom dwelling, temperature variations will inevitably occur. Some rooms, will not get enough airflow, and other rooms will get too much airflow. If motorized dampers are not installed in the system, the only recourse is to manually adjust the air vents in each room. Manual setting of the room vents may solve the problem, but only temporarily. And when the seasons change, the adjustments may also have to be changed.

PROCEDURES

Thermostats

The thermostat is typically the centerpiece of a residential temperature control system. It acts as a process controller that measures the ambient temperature, compares it to a user-defined set point, and creates an output action to force the two values together. This means that the output action will call for the application of heat when the measured temperature is below the set point value established by the user.

Because the thermostat sensor for this procedure is located downstairs and there is no control of the upstairs, heating and cooling variations will occur. All heating and cooling calls are made to the system by a thermostat that has no information about the temperature upstairs. The result will be an upstairs that is overheated in the winter and undercooled in the summer.

1. Turn on the main ac power switch and supply power to the 24 Vac transformer.

2. Make sure the Zone 1 thermostat is in the *OFF* mode by pressing the **Mode** button on the thermostat's front panel until its LCD displays OFF, as illustrated in Figure 10-1.

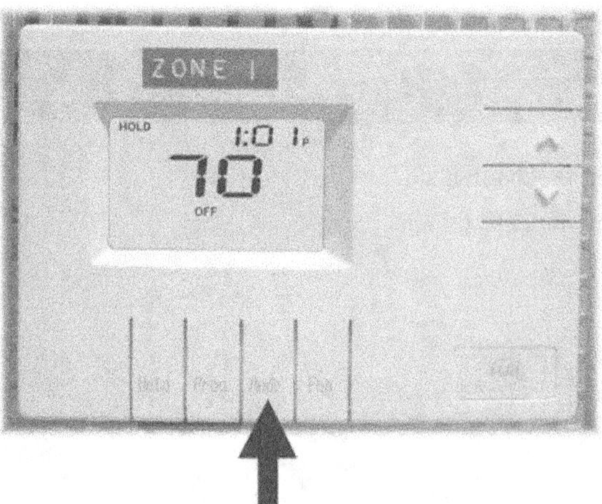

**Figure 10-1:
Thermostat in
OFF Mode**

3. Record the current temperature reading of the wall thermometer for the upstairs environmental box in Table 10-1, column A.

Table 10-1: Current Temperature and Environmental Box Temperature Variations

	Current Temperature	Resulting Temperature Differences
Upstairs Environmental Box Thermometer		
Downstairs Thermostat		
	A	B

4. Record the current temperature reading from the downstairs thermostat in Table 10-1, column A.

5. Put the thermostat into the *COOL* mode by pressing the **Mode** key until COOL is displayed on the unit's LCD display, as shown in Figure 10-2.

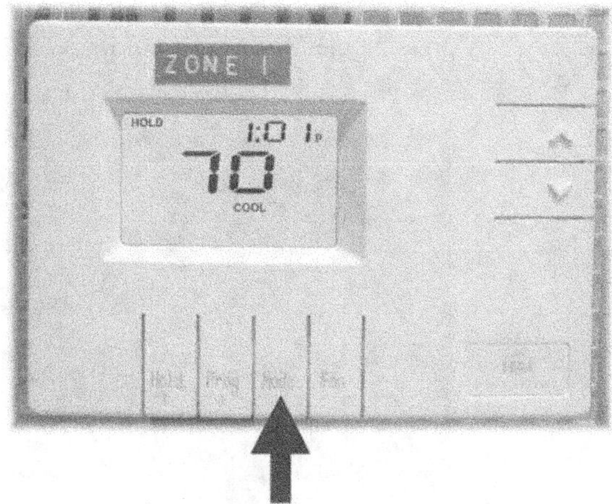

**Figure 10-2:
Thermostat in
COOL Mode**

6. Set the desired temperature the LCD by pressing the (∧) or (∨) keys respectively.

7. Use the downstairs **environmental variable controls** to supply external heat to the bottom room of the residence. See Figure 10-3.

**Figure 10-3:
Environmental
Variable Control Set
to Day/Noon**

8. Allow the downstairs environmental box to heat until the thermostat calls for cooling. Then turn the downstairs environmental variable controls **off** by putting the three-way switch back to the **center position**.

9. Allow the furnace/air handler unit to cool the downstairs room of the residence until the cooling cycle is complete and the air handler shuts itself off.

10. Increase the desired temperature on the thermostat five degrees by pressing the (∧) key on the thermostat.

11. Put the thermostat into *HEAT* mode by pressing the **Mode** key until HEAT is displayed on the LCD display, as illustrated in Figure 10-4.

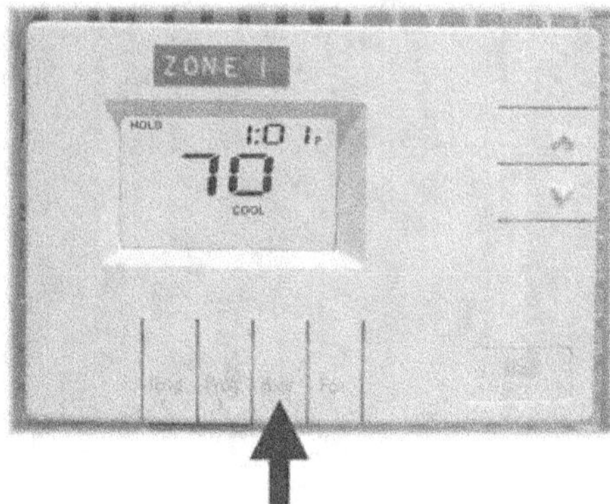

Figure 10-4:
Increasing the Desired
Temperature on the Thermostat

12. Allow the heating cycle to complete. When it shuts off, set the desired temperature back down five degrees.

13. Press the **Mode** button until the thermostat's LCD displays the *OFF* status.

14. Record the temperature from the wall thermometer in the upstairs environmental box in Table 10-1, column B.

15. Record the temperature from the downstairs thermostat in Table 10-1, column B.

16. Turn off the single-pole general-use ac switch and discontinue power to the HVAC system.

Thermostat Connections to the HVAC Controller

(Repeat of Zone 1 for Zone 2)

1. Review the manufacturer's installation and operation manuals for the HVAC controller and thermostat.

2. Use the information supplied by the manufacturer, along with input from Figure 10-5 to complete Table 10-2.

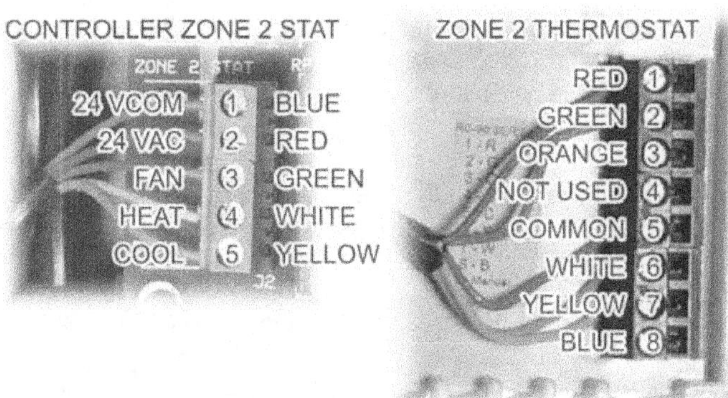

Figure 10-5: Connection Diagram for Zone 2 STAT to Zone 2 Thermostat

Table 10-2: Connection Block Diagram for ZONE 2 STAT to ZONE 2 Thermostat

ZONE 1 STAT		Wire Color from Thermostat	Thermostat 1		Wire Color from Controller
PIN 1	24COM B		PIN 1	1-R	
PIN 2	24VAC R		PIN 2	2-G	
PIN 3	FAN G		PIN 3	3-O	
PIN 4	HEAT W		PIN 4	4-n/u	
PIN 5	COOL Y		PIN 5	5-C	
			PIN 6	6-Y	
			PIN 7	7-W	
			PIN 8	8-B	

3. Have your instructor verify the contents of Table 10-2 for accuracy.

4. Make the wire connections to the HVAC controller's ZONE 2 STAT terminal block.

5. Make the wire connections to the Zone 2 thermostat's terminal block.

Remote Sensor Connections to the Thermostat

(Repeat of Zone 1 for Zone 2)

1. Review the manufacturer's installation and operation manuals for the thermostat and remote temperature sensor.

2. Use the information supplied by the manufacturer, Figure 10-5, and data contained in Table 10-2 to complete the connections between the remote sensor and the thermostat.

3. Use wire nuts to make the connections between the 2-wire shielded cable and the remote sensor connector.

4. Use wire nuts to make the connections between the 2-wire shielded cable and the thermostat's **J4** connector.

5. Attach the **J4** connector to the thermostat.

6. Mate the thermostat to its housing and clip it into place.

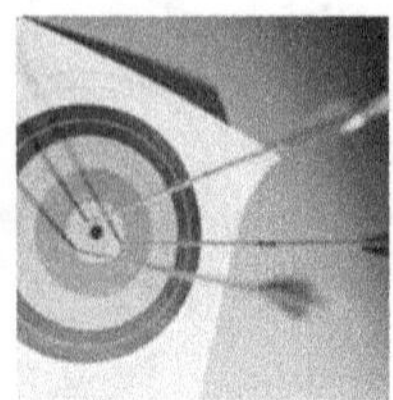

Feedback

LAB QUESTIONS

1. What type of thermostat provides overall control of HVAC support to a structure?

2. How is localized temperature control of the HVAC system maintained in a structure?

3. What happens if a non-zoned HVAC system is installed in a multilevel structure?

4. What is the function of the thermostat in an HVAC system?

Installing and Configuring Multizone HVAC Systems

OBJECTIVES

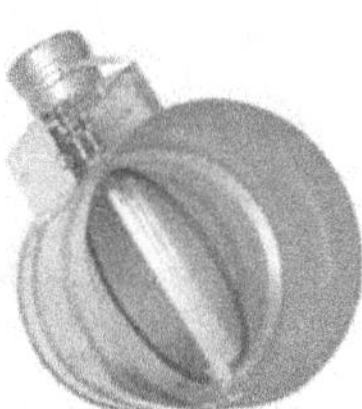

HVAC Systems

1. Identify and describe the function of a motorized damper in an HVAC system.
2. Identify wire connection points between the HVAC controller and the motorized dampers.
3. Recognize the wire color-coding scheme for the HVAC controller and the motorized dampers.
4. Interconnect the parts discussed within this lab procedure.
5. Use the Installer Setup mode.
6. Access and adjust the calibration-offset value for two thermostats.
7. Verify that the thermostats are functioning as independent zones.
8. Verify that the zone dampers are functioning properly.
9. Create a call for the cooling cycle.
10. Create a call for the heating cycle.
11. Simulate temperature differences between rooms caused by environmental warming.
12. Simulate temperature differences between rooms caused by the use of household appliances.
13. Simulate temperature increases of the rooms caused by intense environmental warming.
14. Simulate temperature decreases of the rooms caused by intense environmental cooling.

RESOURCES

1. Marcraft HVAC Experiment Panel and Frame

TOOLS

1. Flat-tip screwdriver, precision head
2. Philips screwdriver, standard 6 inch

DISCUSSION

Damper controls use movable plates located in the distribution ductwork to regulate airflow to different areas of the residence. Typically used in a zoning application, they direct heated or cooled air to the areas that need it most. To do this, they normally limit the flow of heated/cooled air by shutting off the airflow to unoccupied or unused rooms. Due to the increased airflow in the occupied areas, the temperature changes more quickly, resulting in shorter run times for the heating/cooling system.

Certain rooms are habitually warm or cool, regardless of the settings at the thermostat. If some rooms are too hot or cold, try adjusting the dampers in the registers. If the system has them, adjust the dampers in the warm-air ducts.

When a duct run is long the airflow through it may decrease to the point where it is no longer effective. It may then be necessary to install an in-line duct fan.

PROCEDURES

Motorized Damper Wiring

Choose motorized dampers based upon the configuration, style, and size of ductwork. Motorized dampers come in a variety of sizes, but the smallest *standard* round size is 6 inches. Motorized damper types also come in square and rectangular configurations. They are normally placed in the ductwork at a point of airflow divergence. This facilitates redirection of the airflow to the zone(s) that need(s) it most.

The motor wires for the dampers are normally not polarity sensitive. Wiring the dampers is simple, and requires one signal wire pair per damper. The wire type is the same as that used for wiring the thermostat to the controller (18 AWG). The wire termination points should be clearly marked on the controller. The style of controller chosen will determine the number of motorized dampers that can be installed.

Although motorized dampers help to redirect airflow in a structure, it is important to remember that heat rises and cold air sinks in a multilevel structure. If the structure does not have closeable openings between levels, a temperature variation situation will occur. In a single-zone system, the problem will be uneven heating and cooling. But with a multizone system, the HVAC equipment will be overworked trying to compensate for the differences caused by convection.

1. Review the installation and operation manuals for the controller and the motorized dampers.

2. Use the information supplied by the manufacturer, and Figure 11-1, to wire both the Zone 1 and Zone 2 dampers to the controller.

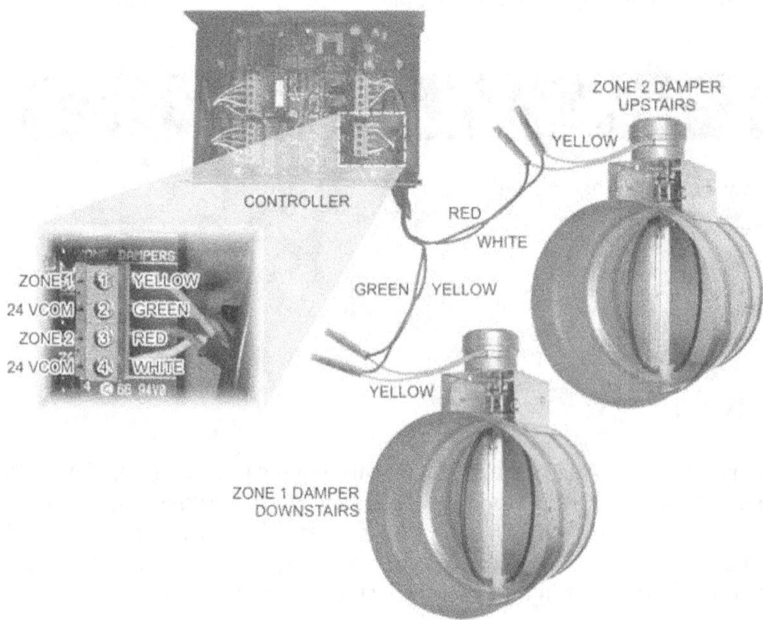

Figure 11-1: Controller and Motorized Damper Wiring

3. Record the wire color configuration of connections to the controller and to the motorized dampers in Table 11-1.

Table 11-1: Motorized Dampers Wire Color Configuration

Wire Color Connections from Zone 1 Damper to Zone 1 Controller Terminals		Wire Color Connections to Motorized Damper from Controller Terminals Zone 1 and 24VCOM (Downstairs)	
Zone 1			
24VCOM			
Wire Color Connections from Zone 2 Damper to Zone 2 Controller Terminals		Wire Color Connections to Motorized Damper from Controller Terminals Zone 2 and 24VCOM (Upstairs)	
Zone 2			
24VCOM			

Thermostat Calibration Offset

HVAC systems require that various components be set for proper operation. The temperature sensor provided by the factory is removed when a remote temperature sensor is installed, and this may cause an inaccurate reading. So it is imperative that the thermostats be calibrated to the actual temperature of the room where the remote sensors are installed.

1. Turn the Marcraft HVAC Experiment Panel on.

2. Ensure that both thermostats are in the **OFF** mode by pressing the **Mode** button until both thermostats' LCD displays show OFF, as shown in Figure 11-2.

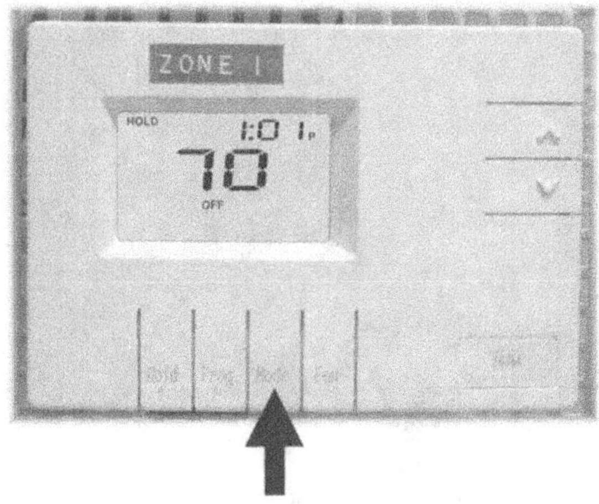

Figure 11-2: Mode Button

3. Determine whether the remote temperature sensors are calibrated to the reading of the external thermometer.

4. Record the temperature reading of the external thermometer in column A of Table 11-2.

Table 11-2: Temperature Readings and Calibration Offsets

	Temperature	External – Zone	Value	Calibration Offset + Value	New Reading for Calibration Offset
Zone 1					
Zone 2					
External Thermometer					
	A	B	C	D	E

5. Record the temperature reading of the Zone 1 (downstairs) thermostat in column A of Table 11-2.

6. Record the temperature reading of the Zone 2 (upstairs) thermostat in column A of Table 11-2.

7. Use columns B−E of Table 11-2 to calculate the calibration offset values for both thermostats, by completing the following steps.

Example 1: If the external thermometer reads 70 and the Zone 1 thermostat reads 72, then the calibration offset value will need to be decreased by 2. $30 - 2 = 28$. Set the value of item 04 from 30 to 28.

Example 2: If the external thermometer reads 70 and the Zone 2 thermostat reads 70, then the calibration offset value will not need to be decreased. $30 - 0 = 30$. Leave this value alone.

8. Enter the Installer Setup for the Zone 1 (downstairs) thermostat by pressing the **Prog** (>) key three times, then press the **Fan** key. See Figure 11-3.

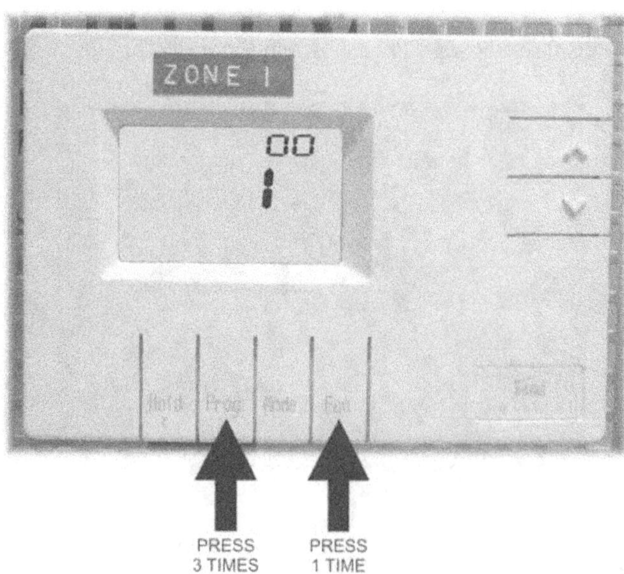

Figure 11-3: Prog (>) and Fan Keys

NOTE: The thermostat will automatically exit Setup mode after 20 seconds of no key activity occurs.

9. Advance to the *Calibration Offset* item by pressing the **Prog** (>) key until the item number reads **04**, as illustrated in Figure 11-4.

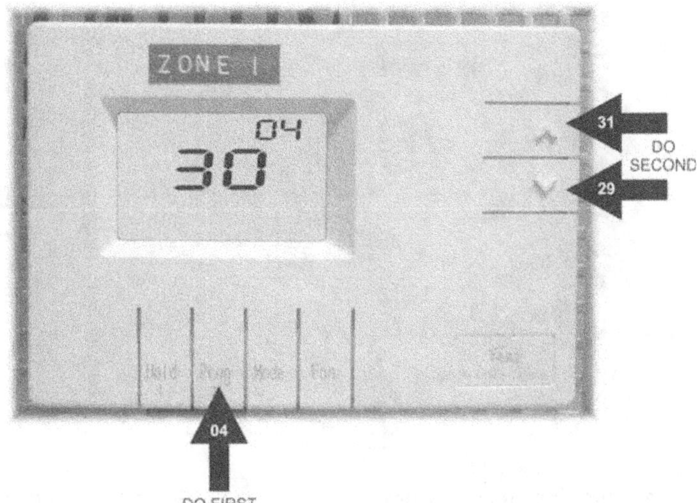

**Figure 11-4:
Item Number
Reading 04**

10. Increase or decrease the value of item 04 to match the value calculated in column E of Table 11-7, by pressing the (∧) or (∨) keys respectively.

11. Exit the Installer Setup by pressing the **Fan** key.

12. Repeat steps 8–11 for the Zone 2 (upstairs) thermometer.

13. Turn off the Marcraft HVAC Experiment Panel.

Spring Morning Simulation

By zone programming the HVAC system, you can properly cool the upper level of a two-story residence without overcooling the downstairs. When a zone calls for cooling, the controller shuts the damper for the non-calling zone. This will redirect cooling to the zone calling for it. When the cooling cycle is complete, the damper for the non-calling zone will reopen. If the closed zone needs cooling during the cooling process, the controller will simply open the damper during the cooling cycle.

NOTE: The thermostats should be in the OFF mode.

1. Turn the Marcraft HVAC Experiment Panel on.

2. Record the current temperatures from the Zone 1 and Zone 2 thermostats in column A of Table 11-3.

Table 11-3: Current Temperature/Desired Temperature

	Current Temperature	Desired Temperature
Zone 1		
Zone 2		
	A	**B**

3. Press the **Mode** key for the Zone 1 thermostat until the LCD display reads **AUTO**, as shown in Figure 11-5.

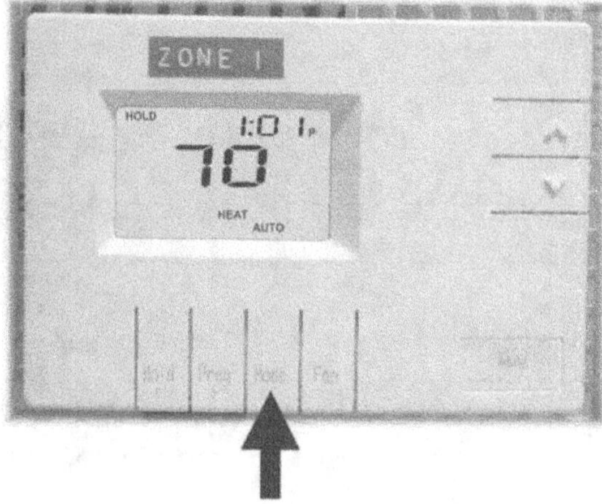

**Figure 11-5:
Changing the Thermostat
Function to AUTO**

4. Set the DESIRED temperature for Zone 1 by pressing the (∧) or (∨) keys on the thermostat to the current temperature recorded for Zone 1 in column A of Table 11-8. See Figure 11-6.

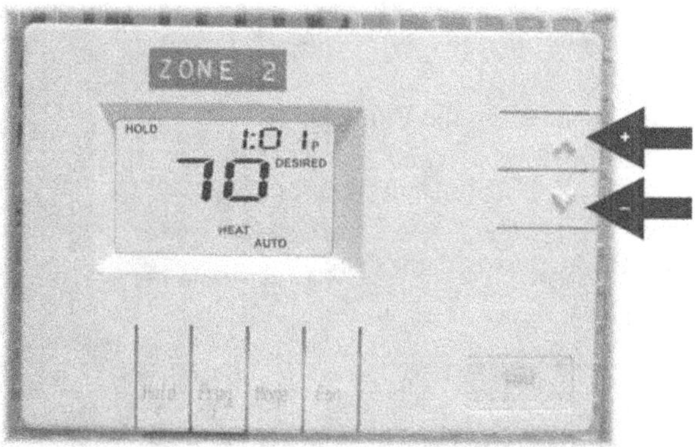

**Figure 11-6:
Increasing or Decreasing
the DESIRED
Temperature Reading**

5. Record the DESIRED temperature reading for the Zone 1 thermostat in column B of Table 11-3.

6. Repeat steps 3 through 5 for the Zone 2 (upstairs) thermostat. This action places the system in a neutral mode, where no cooling or heating activities are required.

**Figure 11-7: Applying
External Heating to
Zone 2 with the
Upstairs Environmental
Variable Controls**

7. Select the **Day** setting on the *Upstairs Environmental Variable* controls and turn the bulb brightness control knob to half way between dawn and noon, as shown in Figure 11-7. If warming occurs too slowly, increase the bulb intensity as needed.

NOTE: The approximate time required for heating the room from 70 degrees to the controller's activation with the bulb set at half power is 25 minutes.

NOTE: This will simulate the natural heating caused by the sun as the roof is warmed.

8. Allow the heating to continue until the controller turns the air handler on, then move the *Day* switch for Zone 2 to **OFF**.

9. Record the current temperature readings for both thermostats in column A of Table 11-4.

Table 11-4: Temperature That Fan Comes On and Turns Off

	Current Temperature	Temperature When Fan Turns Off	Current Temperature	Temp When Fan Turns Off
Zone 1				
Zone 2				
	A	B	C	D

10. Verify that Zone 2 is calling for cooling. See Figure 11-8.

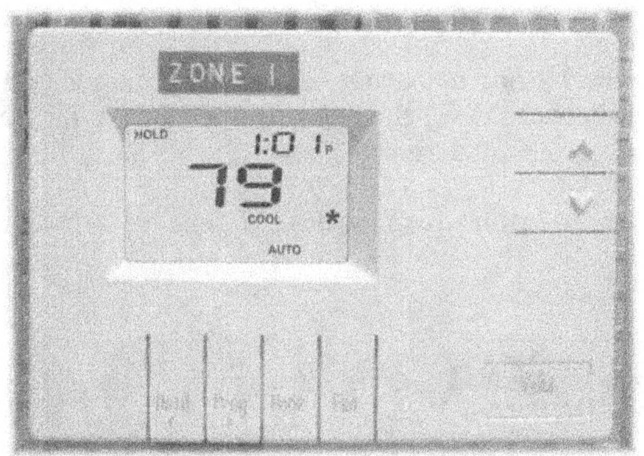

Figure 11-8: Thermostat Displaying COOL with a (★) to the Right

11. Record the state of the zone dampers in columns A and B of Table 11-5.

Table 11-5: Zone Damper State

	Open	Closed	Open	Closed	Open	Closed
Zone 1						
Zone 2						
	A	B	C	D	E	F

12. Record the temperature at which the air handler shuts off for Zone 2 in column B of Table 11-4.

13. Turn both thermostats off by pressing the **Mode** key until both LCDs display **OFF**.

14. Turn off the Marcraft HVAC Experiment Panel.

Appliance Warming Simulation

Placing thermostats near heat sources such as warm air ducts, gas or electric ranges, clothes dryers, fireplaces, and ovens will guarantee the improper operation of any HVAC system. Rooms some distance from these devices will feel uncomfortably cool. In addition, locating a thermostat near less evident sources of heat such as lamps and electronic equipment can also become a problem.

Because outside walls conduct some degree of heat and cold originating from outside of the house, these temperature variations will adversely affect the thermostat's accuracy. In order to avoid this, mount thermostats on interior or partitioning walls.

1. Turn the Marcraft HVAC Experiment Panel on.

2. Put both thermostats into **AUTO** mode.

3. Set the DESIRED temperatures for both thermostats to the current remote sensor readings of the thermostats.

4. Select the **Day** setting on the *Downstairs Environmental Variable* controls and turn the bulb brightness control knob to halfway between dawn and noon, as illustrated in Figure 11-9. If warming occurs too slowly, increase the bulb intensity as needed.

NOTE: This will simulate the heating caused by the use of major appliances.

Figure 11-9: Applying External Heating to Zone 1 with the Downstairs Environmental Variable Controls

5. Allow the heating to continue until the controller turns the air handler on, and then turn off the *Day* switch for Zone 1.

6. Record the current temperature readings for both thermostats in column C of Table 11-4.

7. Verify that the Zone 1 thermostat is calling for cooling, as shown in Figure 11-10.

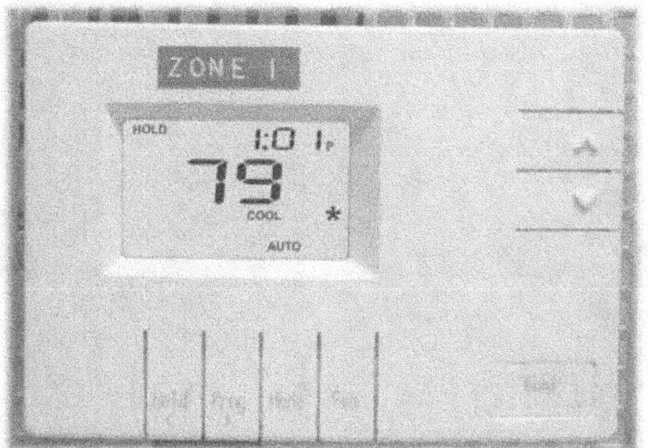

**Figure 11-10:
Thermostat LCD
Displaying COOL
with a (★) to the Right**

8. Record the state of the zone dampers in columns C and D of Table 11-5.

9. Record the temperature at which the air handler for Zone 1 turns off in column D of Table 11-4.

10. Turn both thermostats to the *OFF* condition by pressing the **Mode** key until their LCD displays show **OFF**.

11. Turn off the Marcraft HVAC Experiment Panel.

Summer Day Simulation

Technicians should remember that during the summer, every degree that the air conditioning thermostat is raised will result in a 5% decrease in cooling energy consumption. Likewise during the winter, every degree that the thermostat is lowered will result in a 3% decrease in heating energy consumption. Thermostats that regulate unoccupied rooms in the home can save substantial amounts of energy by operating at a setting of 55°F in the winter, and above 80°F in the summer.

During warm weather, most homeowners know enough about their HVAC equipment to switch the thermostat to the cool setting, to set the fan to its automatic mode, and to adjust the temperature setting low enough to operate the AC section. The majority of users also understand the importance of keeping the system filters clean. Beyond these basic measures, however, common residential HVAC problems may go unrecognized.

1. Turn the Marcraft HVAC Experiment Panel on.

2. Put both thermostats into **AUTO** mode.

3. Set the DESIRED temperatures for both thermostats to the current remote sensor readings of the thermostats.

4. Select the **Day** setting on the *Downstairs* and *Upstairs Environmental Variable* controls and turn the bulb brightness control knobs to **noon**, as illustrated in Figure 11-11. Warming should occur rapidly.

NOTE: This will simulate the heating caused by the sun on a hot summer day.

Figure 11-11: Applying Full External Heating to Zones 1 and 2 with the Downstairs and Upstairs Environmental Variable Controls

5. Allow the heating to continue until the controller turns the air handler on, then turn off the **Day** switch for Zones 1 and 2 and record the temperature readings for both thermostats in column A of Table 11-6.

Table 11-6: Temperature at which Fan Comes On and Turns Off

	Current Temperature	Temperature When Fan Turns Off
Zone 1		
Zone 2		
	A	**B**

6. Verify that the Zone 1 and Zone 2 thermostats are calling for cooling.

7. Record the state of the zone dampers in columns E and F of Table 11-5.

8. Record the temperature at which the air handler turns off for Zone 1 and Zone 2 in column B of Table 11-6.

9. Turn both thermostats off by pressing the **Mode** key until both LCD displays show **OFF**.

10. Turn off the Marcraft HVAC Experiment Panel.

Winter Day Simulation

Normally, residential temperature control is performed in an On/Off manner. When the room temperature drops below the thermostat's set point, the connection is closed and power is applied to the heating elements. Likewise, when the temperature in the vicinity of the thermostat rises above the setting on the thermostat, an electrical connection in the thermostat opens and the controller removes power from the heating elements, causing them to cool down.

1. Turn the Marcraft HVAC Experiment Panel on.

2. Put both thermostats into **AUTO** mode.

3. Set the DESIRED temperatures for both thermostats 5 degrees higher than the current remote sensor readings of the thermostats. Record these new DESIRED values for Zone 1 and Zone 2 in column A of Table 11-7.

Table 11-7: Winter Day Simulation Temperature Reading

	Temperature Desired	Current Temperature	Temperature When Heating Starts
Zone 1			
Zone 2			
	A	B	C

4. Allow the controller to respond to the request for heating.

5. Record the temperature at which the furnace shuts off for Zone 1 and Zone 2 in column B of Table 11-7.

6. Select the **Night** and **Midnight** settings on both of the Environmental Variable controls. See Figure 11-12.

Figure 11-12: Environmental Variable Controls Set To Night/Midnight

7. Allow the cooling to continue until the controller calls for heat and then turn off the **Night** switches for Zone 1 and Zone 2.

8. Record the current temperatures displayed on the thermostats when heat is called for a second time in column C of Table 11-7.

9. Turn both thermostats off by pressing the **Mode** key until both LCD displays show **OFF**.

10. Turn off the Marcraft HVAC Experiment Panel.

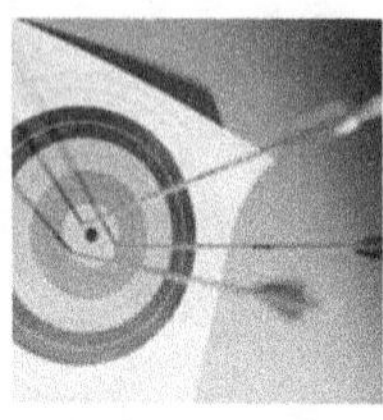

Feedback

LAB QUESTIONS

1. What is the function of a motorized damper in an HVAC system?

2. If a room is habitually warm or cold, regardless of the settings at the thermostat, what measure (other than motorized dampers) can be taken to compensate for this?

3. What is the smallest standard size round motorized damper?

4. Why should a thermostat not be placed near appliances that produce heat?

5. What is the importance of keeping the system filters clean?

Multizone HVAC Systems Research

OBJECTIVES

1. Research various zoned HVAC components.
2. Locate applicable local, state, and federal regulations.

**HVAC
Systems**

RESOURCES

1. Books
2. Newspapers
3. Magazines
4. PC system with graphic capture software and Internet access

DISCUSSION

Gathering meaningful information is not as difficult as it once was, because modern students conducting serious research have numerous sources of information available to them. Yet, the ability to conduct research remains one of the most important skills required for successful scholarship. Granted, traditional researching methods continue to require the use of a card file from which various books, newspapers, and magazines may be selected for study. But the Internet has opened up new opportunities for conducting information searches about almost any subject of interest. To take advantage of this recent development, modern libraries have computer terminals for conducting Internet searches.

Drowning in an ocean of unwanted information is a real danger for inquisitive minds when searching the Internet without basic researching skills. Students must be able to locate the desired information quickly, without becoming ensnared in time-wasting searches through mountains of online information. Thousands of individual files are being added to the Internet even as you read this paragraph, and most of them are not catalogued for easy identification and retrieval. Among the many Internet data delivery services available are electronic mailings, file transfers, interest group memberships, interactive collaborations, and multimedia displays.

It's true that World Wide Web (WWW) browsers include most of the Internet protocols required to effectively manipulate the data services previously mentioned. But the addresses of Internet sites frequently change, or the sites disappear altogether. When this happens, you may see a screen similar to the one shown in Figure 12-1. This screen testifies to the basic instability of the Internet! To overcome this, conduct your Internet research by copying and storing any information (text and graphics) that appears useful to the project at hand. That way, if the Internet site from which the material was obtained ever disappears, the information itself is still available to you.

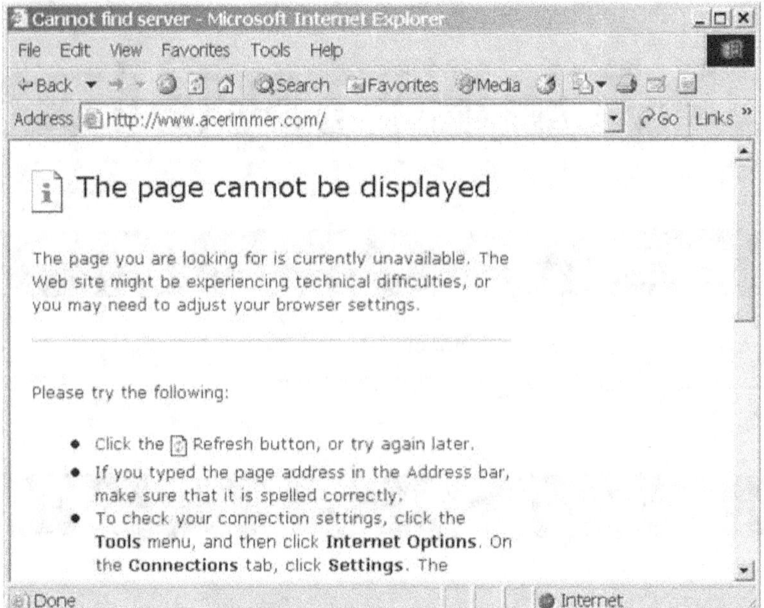

Figure 12-1: Internet Site or Page Cannot be Displayed

Direct computer transfer of certain materials from proprietary or protected Internet sites sometimes is not possible, especially information from product vendors. In these cases, it may be necessary to type the relevant text, by hand, to a word processor for data storage. Graphics may have to be captured using a third-party software tool, and later converted into a usable picture format.

When conducting research from the Internet, or from more traditional sources, you must verify the integrity of the information. Book, newspaper, magazine, or Internet articles may look interesting, but are they accurate and truthful? Veracity is strengthened by obtaining identical information from a variety of sources!

HVAC Systems

PROCEDURES

Because efficiency is so important in the design and operation of HVAC, your research should concentrate primarily on components that make up a zoned system. Constraints on the availability of energy supplies and on their affordability make the use of HVAC zoning systems imperative.

To make your allotted research time more productive, concentrate on information about the zoned HVAC components in the product areas listed below. This research task will be of benefit when weighing factors for a particular installation, including the size of the residence, its layout, and the particular requirements of the client. Information gathered here could be used to consider and compare specific pieces of equipment, their system capabilities, available vendors, and estimated costs.

1. Examine the following list. It identifies HVAC components with which a RESI Environmental Control technician will most likely be working.

- Air handlers
- Communicating thermostats
- Communications cable
- Computers
- Condensers
- Dampers
- Ducts
- Furnaces
- Heat pumps

- Motorized dampers
- Networking devices
- Radiators
- Sensors
- Solar collectors
- Static pressure probes
- Termination equipment
- X-10 controllers
- Zoning/distribution panels

2. Use Tables 12-1 through 12-18 to organize the specified details about the HVAC components listed above. Try to locate more than one vendor for each selected product.

Table 12-1: Air Handlers

Model	Description	Vendor	Cost

Table 12-2: Communicating Thermostats

Model	Description	Vendor	Cost

Table 12-3: Communications Cable

Model	Description	Vendor	Cost

Table 12-4: Computers

Model	Description	Vendor	Cost

Table 12-5: Condensers

Model	Description	Vendor	Cost

Table 12-6: Dampers

Model	Description	Vendor	Cost

Table 12-7: Ducts

Model	Description	Vendor	Cost

Table 12-8: Furnaces

Model	Description	Vendor	Cost

Table 12-9: Heat Pumps

Model	Description	Vendor	Cost

Table 12-10: Motorized Dampers

Model	Description	Vendor	Cost

Table 12-11: Networking Devices

Model	Description	Vendor	Cost

Table 12-12: Radiators

Model	Description	Vendor	Cost

Table 12-13: Sensors

Model	Description	Vendor	Cost

Table 12-14: Solar Collectors

Model	Description	Vendor	Cost

Table 12-15: Static Pressure Probes

Model	Description	Vendor	Cost

Table 12-16: Termination Equipment

Model	Description	Vendor	Cost

Table 12-17: X-10 Controllers

Model	Description	Vendor	Cost

Table 12-18: Zoning/Distribution Panels

Model	Description	Vendor	Cost

Applicable Regulations

A completed HVAC installation says a great deal about the skill and awareness of the RESI Environmental Control technician(s) who installed it. Failure to consult the local, state, or federal regulating bodies having jurisdiction over residential or commercial HVAC systems could say something you would rather not have repeated.

Local

Strict adherence to federal and state regulations will often result in acceptable HVAC inspections being conducted, but this does not guarantee protection against some finicky local code. Awareness of special local building ordinances must also be taken into account, before installation of a multizone HVAC system begins! These codes usually involve zoning requirements, subdivision limitations, mobile home restrictions, or flood plain ordinances. The local planning commission is a good source for this type of information because these officials are most responsible for controlling development in the local community, including rezoning, building codes, HVAC installations, and the exterior location of system components.

Contact your local planning commission and gather information about requirements for HVAC installations and/or equipment not specifically addressed by state or federal codes. Although an RESI Environmental Control technician has experience with HVAC installations in one county, this doesn't ensure that he or she will know system requirements in another. Unique guidelines could be enforced by either city or county agencies. For example, state watershed designations could place unique restrictions on the local use of certain utilities. Alternately, a shortage of power generating capacity will place limitations on the use of specific types of equipment.

1. Use Table 12-19 to organize specific details about the HVAC installations and/or equipment strictly controlled by your local codes. Try to locate more than one source of information for each consideration or restriction.

Table 12-19: Local HVAC Codes or Regulations

Information Source	Specified Equipment	Restrictions	Special Considerations

State

State-sponsored HVAC regulations normally center on setting and collecting license fees, and ascertaining that licensed HVAC installers are properly insured. They also establish minimum codes and standards for the installation, alteration, remodeling, maintenance, and repair of heating, ventilation, air conditioning, and refrigeration systems. State licensing boards investigate complaints against both licensees and those involved in unlicensed practice. They have the authority to reprimand, suspend, or revoke a valid HVAC license. Under certain circumstances they may also assess a civil monetary penalty when severe violations of the applicable regulations have been uncovered. Information concerning requirements for licensed HVAC technicians in your state can be found by browsing its government Web pages.

Remember that an HVAC technician licensed in one state, will not be able to legally install an HVAC system in another. Licenses must be obtained for all states in which installation work is undertaken.

2. Go to your state's information page and locate HVAC regulations for contractors and installers. Use Table 12-20 to organize specific details about insurance requirements strictly controlled by your state's codes and agencies.

Table 12-20: State HVAC Insurance Requirements

Controlling Agency	Specified Equipment	Regulation ID	Limited Minimums

3. From your state's information page on HVAC regulations for contractors and installers, use Table 12-21 to organize specific details about license requirements strictly controlled by your state's codes and agencies.

Table 12-21: State HVAC License Requirements

Controlling Agency	Specified License	Experience Required	Minimum Exam Score

Federal

Manufacturers and consumers around the world recognize various technical bodies that have the authority to determine the suitability of one product over another. These bodies have defined and enforced standards in order to attain some measure of control over the recent explosion of technical development.

The major organizations listed here can be consulted to obtain published standards that determine the acceptability of both products and labor involved in HVAC system installations and operations. Charged with the necessity of determining the suitability of one product over another, these organizations have published standards to guide the RESI Environmental Control technician in the selection of appropriate equipment for a particular
installation.

National Electrical Code

The National Electrical Code, sponsored by the National Fire Protection Agency, contains information upon which all aspiring electricians must be tested before obtaining their licenses. New users of this document should use its index, notwithstanding the fact that most electricians can easily locate any required information in the code. This is so only due to the fact that licensed electricians have spent a great deal of time studying it.

Most states require a permit and an inspection for performing electrical work, and the National Electrical Code is considered to be merely a guideline. Following the NEC cannot guarantee safe electrical installations, but it is the best guide available. Still, an electrician should use his or her own judgment given the many applications available for specific wiring jobs. The best approach is to perform all work above the minimum safety standards set forth by the NEC.

Before performing any HVAC wiring work, contact the local city or town wire inspector. Each state differs slightly in its requirements for inspection and code compliance. Under these circumstances, the local wire inspector is "The Authority Having Jurisdiction" in most locations. He or she is responsible for rule interpretation and for code enforcement.

4. Identify the local wire inspector in your area. Record the information specified in Table 12-22 to organize the details about your local wire inspector.

Table 12-22: The Authority Having Jurisdiction

Business Name	Personal Name	Street Address	Telephone Number	Web Address

TIA/EIA Standards

Standards created by the Telecommunications Industry Association (TIA) and the Electronic Industries Alliance (EIA) are certified by the American National Standards Institute (ANSI). In addition, Telecommunications Systems Bulletins (TSBs) are addenda to, or explanatory comments about, either an industry standard or an interim standard. They are often integrated into forthcoming standard revisions.

5. Conduct an Internet search for any ANSI/TIA/EIA standards pertinent to residential HVAC control and management systems, and record information about them in Table 12-23.

Table 12-23: ANSI/TIA/EIA Residential HVAC Standards

Standard ID	Standard Name	Standard Description	Year Established

IEEE Standards

The Institute of Electrical and Electronics Engineers, Inc. (IEEE) is a nonprofit, technical professional association of more than 377,000 individual members from more than 150 different countries. Because of the size and scope of its membership, the IEEE is a leading authority in various technical areas including computer engineering, biomedical technology, telecommunications, electric power, aerospace, and consumer electronics.

6. Conduct an Internet search for any IEEE standards pertinent to residential HVAC control and management systems, and record information about them in Table 12-24.

Table 12-24: IEEE Residential HVAC Standards

Standard ID	Standard Name	Standard Description	Year Established

Underwriters Laboratories Inc. (UL)

Underwriters Laboratories Inc. (UL) is an independent, not-for-profit, product safety testing and certification organization and has tested products for public safety for more than 100 years. Billions of UL marks are applied to products from all over the world. Founded in 1894, UL holds an undisputed reputation as the leader in both U.S. and international product safety and certification. It is one of the most recognized, reputable conformity assessment providers in the world.

Currently, UL services also include helping companies achieve global acceptance for their product, whether it is an electrical device, a programmable system, or a quality process. Listing various manufacturers for specific types of installations and equipment, UL tests equipment for specific uses, and judges installations against the governing standards.

7. Use an Internet search to locate several UL standards that have been incorporated by their manufacturers into some of the HVAC components listed earlier. Record the information you find about applicable UL standards in Table 12-25. Feel free to extend the table if necessary.

Table 12-25: UL HVAC Standards Incorporated by Manufacturers

HVAC Component	Manufacturer	Standard ID	Specified Requirements

National Fire Protection Association (NFPA)

The National Fire Protection Association (NFPA) was founded in 1896 in an effort to reduce the worldwide burden of fire and other hazards on the quality of life. Since then, it has developed and advocated a scientifically based consensus of codes, standards, research, training, and education. Today, the NFPA, located in Quincy, Massachusetts, is an international, nonprofit membership organization that has more than 75,000 members representing nearly 100 nations, and more than 320 employees around the world. It sets installation guidelines for fire equipment, and is the world's leading advocate of fire prevention. As an authoritative public safety source, it holds classes to instruct various installers on how to follow the guidelines.

8. Use an Internet search to locate several NFPA standards that have been incorporated by their manufacturers into some of the HVAC components listed earlier. Record the information you find about applicable NFPA standards in Table 12-26. Feel free to extend the table if necessary.

Table 12-26: NFPA HVAC Standards Incorporated by Manufacturers

HVAC Component	Manufacturer	Standard ID	Specified Requirements

LAB QUESTIONS

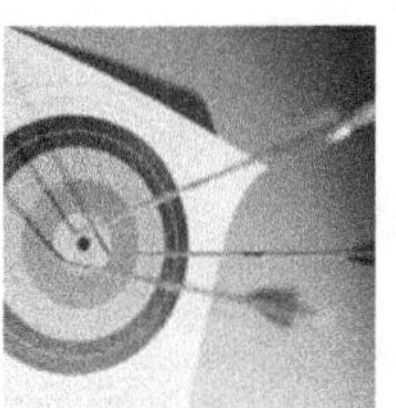

Feedback

1. Why should your research concentrate primarily on components making up a zoned HVAC system?

2. Why is identical information from a variety of sources an important Internet research consideration?

3. What is the benefit of performing research on specific pieces of HVAC equipment, their system capabilities, available vendors, and estimated costs?

4. Why should you take special local building ordinances into account before installation begins?

5. Who is responsible for rule interpretation and code enforcement at the local level?

Designing Residential HVAC Systems

OBJECTIVES

1. Design a cost-efficient home environmental system that will support existing and future needs.
2. Identify the terminology unique to the HVAC trade.
3. List the system components.
4. Describe the benefits of zoning.
5. Recommend a distribution system.
6. Advise on the cabling and thermostat requirements.
7. List compliance standards.
8. Identify suitable and unsuitable locations for thermostats.
9. Recommend an IP address range for the control network.
10. Specify an interface for the zone controller.
11. Identify a piece of equipment that improves airflow within the system.
12. Identify a piece of test equipment for monitoring airflow in the duct system.
13. Recommend some maintenance procedures for the system.

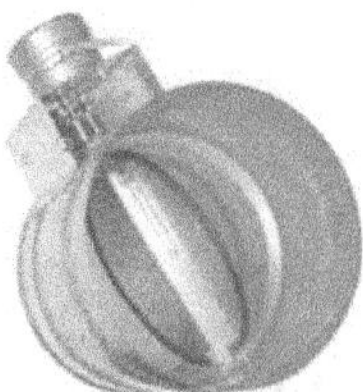

**HVAC
Systems**

RESOURCES

1. Marcraft's *RESI Environmental Control Endorsement* textbook
2. Pencil

DISCUSSION

Read the following scenario:

Digital Automation Ltd is a technology service and products supplier that serves the home integration needs of the building industry in Silicon Valley. The company is currently working on an extensive project that will integrate many advanced electronic components into one, easy-to-use automation system. A licensed HVAC contractor will install the climate control element of the project and you will assist the contractor with zoning specifications for the system, and interfacing with an external automation system.

The system will include one HVAC zone controller, wall display thermostat units, and a damper unit for each zone.

Table 13-1 includes general requirements that were gathered from interviews and written documentation.

Table 13-1: Connection Block Diagram For HVAC System

Requirement Category	Details
Zoning	Two separate zones are to be set up — upstairs and downstairs.
Integration	The HVAC equipment must be integrated into the home automation environment.
Programming	The client should be able to easily program selectable climate modes for different personal tastes, times of day, seasons, vacations, and work schedules.

PROCEDURES

HVAC
Systems

Planning

1. Match each of the following terms with its definition in Table 12-2:

    - Damper control
    - Air handler
    - Condenser
    - Furnace
    - Control panel
    - Thermostat

Table 13-2: HVAC Terminology

Term	Definition
	Used to control temperature
	Used to move the heated or cooled air through the residential ductwork
	Used to direct air to the areas that need it most
	The main generator of heat for the house
	Used to lower the temperature of the ambient air
	Used to manage the hot and cold air demands of the various zones

2. The new zone control system consists of:

- a control panel
- a number of zone dampers
- two thermostats

Indicate where each of the listed items is used in the installation by writing its name on the diagram shown in Figure 13-1.

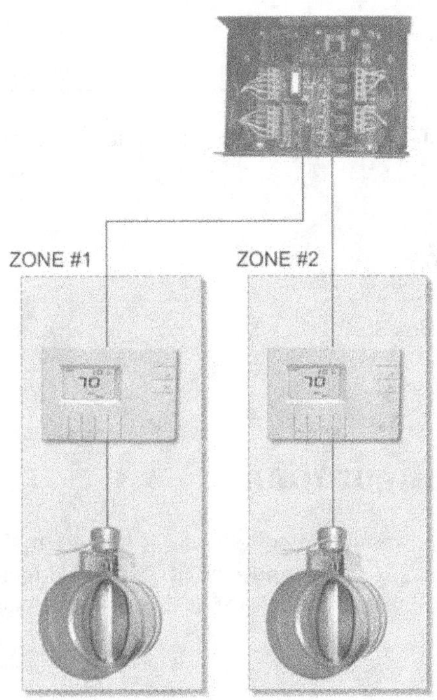

Figure 13-1: Installation Drawing Diagram

3. Your client has requested that the new climate control system be zoned. Which of the following statements about zoning do you think are true?

    ____ a. It improves temperature control.
    ____ b. It reduces fuel costs.
    ____ c. It decreases fuel efficiency.
    ____ d. It improves comfort level.

4. You will construct a distribution system to efficiently distribute the heat from the central furnace in the basement to the various rooms. Which of the following methods of distribution is the most wasteful?

    ____ a. Forced air
    ____ b. Water circulation
    ____ c. Electricity

Explain your answer:

____

____

5. Your main function is to work with the HVAC contractor to coordinate the installation of the HVAC system. What type of cable should you use to interconnect the thermostats with the main zone controller?

____

____

6. To save energy, the new system will heat the upstairs bedrooms automatically at night, while turning off the living area downstairs. What type of thermostat should be installed to provide this type of functionality?

____

7. Which of the following ANSI/TIA/EIA standards pertain to residential HVAC control and management systems?

   ____a. EIA 232F
   ____b. TIA/EIA 570A
   ____c. TIA/EIA 644A
   ____d. EIA 600 CEBus SET

Installation and Configuration

1. A few different types of thermostats are required for this project. In this exercise you will recommend appropriate locations for installing the thermostats. In Table 13-3 identify which of the locations are acceptable (yes) and which are unacceptable (no). If the location is unacceptable as written, explain why.

Table 13-3: Acceptable Locations for Thermostat Installations

Location Guidelines	Yes	No	Reasons
The thermostats for the downstairs zone should be mounted on the exterior wall.			
The thermostats for the downstairs zone should be located close to the front door.			
All thermostats are to be mounted 2 feet (24 inches) above the floor.			

2. On the downstairs floor plan of the house, shown in Figure 13-2, identify a room in the house where thermostats should NOT be located.

____

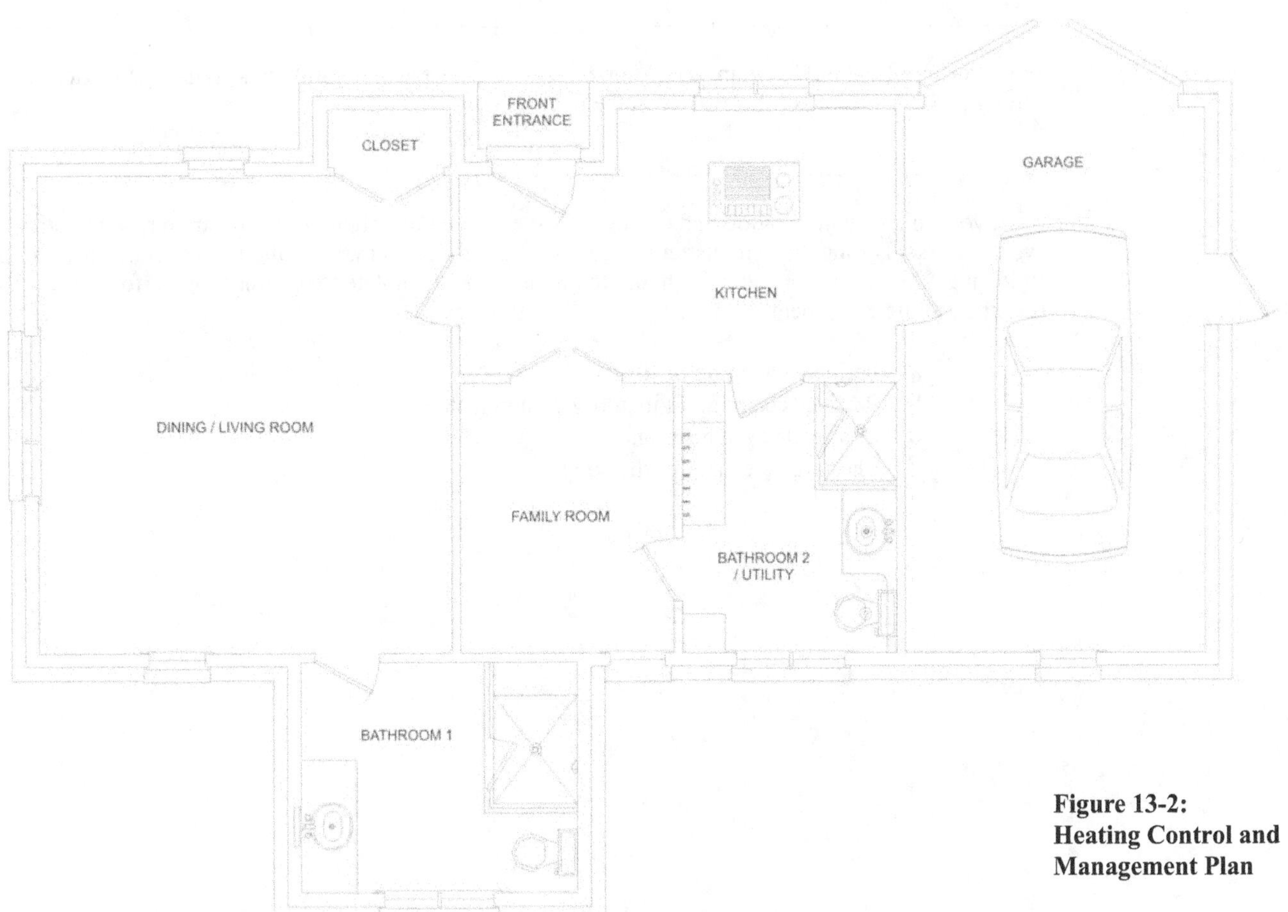

Figure 13-2:
Heating Control and
Management Plan

3. The new HVAC control equipment uses a bridging device to connect with the existing Ethernet-based home network. The bridge must be configured with a unique private IP address. As the home systems integrator, it is your responsibility to suggest an appropriate IP address. Which of the following IP network address ranges may be used for interfacing with the existing home network?

    ____a. 170.16.0.0
    ____b. 10.0.0.0
    ____c. 192.168.0.0
    ____d. 172.16.0.0

4. Combining HVAC equipment with home automation gear is a requirement. The interface supported by the controller should be based on which of the following protocols?

    ____a. USB
    ____b. IEEE 1394
    ____c. X.21
    ____d. RS-232

5. The duct run is so long that airflow through it decreases to the point where it is no longer effective. What piece of equipment should you install in the duct to improve this situation?

    ____

    ____

6. While testing the HVAC system you experience intermittent airflow problems. What test device should you use to monitor the airflow in the duct system?

    ____

7. After you finish the installation and testing, you meet with your client to demonstrate how the new system works. During the demonstration, you provide advice on maintaining the system. Which of the following maintenance procedures should the homeowner complete to optimize the performance of the air handling equipment?

    _____a. Periodic changing of filters
    _____b. Periodic changing of memory components
    _____c. Lubricating fan bearings
    _____d. Thorough cleaning of the coils

Integration Panel's HVAC Connections

OBJECTIVES

1. Identify and describe the function of communicating thermostats and components.
2. Identify connection points between the communicating thermostats and the Integration Controller Panel.
3. Interconnect the parts discussed within this Lab Procedure.
4. Establish a communications link between the communicating thermostats and the controller.
5. Test the integration controller's ability to change the temperature in the two zones.

HVAC Systems

RESOURCES

1. Marcraft HVAC Experiment Panel and Frame
2. Marcraft Integration Panel and Frame
3. 4-conductor shielded cable (length dependent upon the location of the HVAC panel and Integration Controller Panel)
4. Three wire nuts

TOOLS

1. Cable stripper
2. Philips screwdriver
3. Flat-tip screwdriver

DISCUSSION

Various manufacturers have developed access systems that enable the homeowner to control the residential HVAC system remotely through telephone operations and X-10 controllers. Truly convenient HVAC control can be achieved by including a remote access system that will enable the owner to make changes to the HVAC system even when away from the residence. Thermostats can be grouped into two major types, legacy (traditional) thermostats, or intelligent thermostats. Intelligent thermostats are also known as communicating thermostats. They have found wide application in modern HVAC systems. Operating similarly to a conventional thermostat, they are additionally capable of being controlled, either locally or remotely, from a home automation system. The intelligent thermostat may be integrated into a home automation system or may be operated independently with the use of a proprietary system controller.

After altering the upstairs thermostat's address, you will check to be sure that the address of the downstairs thermostat differs from the one assigned to the upstairs unit. Then, you will wire the thermostat to the Omni's controller. After naming and testing the thermostat's modes, you will perform heating and cooling operations from the Omni controller.

**HVAC
Systems**

PROCEDURES

Changing the Thermostat's Address

1. Turn on the HVAC system.

2. On the upstairs thermostat, press the **Prog** key three times and the **Fan** key once to enter *Installer* mode.

3. Use the **up arrow** to increase the value of field 00 from the manufacturer's default of 1 to **2**.

4. On the downstairs thermostat, press the **Prog** key three times and the **Fan** key once to enter *Installer* mode.

5. Make sure the value of field 00 is set to **1**.

6. Put both thermostats in the **OFF** mode.

7. Turn **off** the HVAC system power.

Wiring the Thermostats

1. Refer to the installation manuals for both the controller and the communicating thermostats while performing this lab procedure.

2. Turn **off** the power to the HVAC system, if it isn't already.

3. Refer to Figure 14-1, while making the connections from the thermostats to the controller.

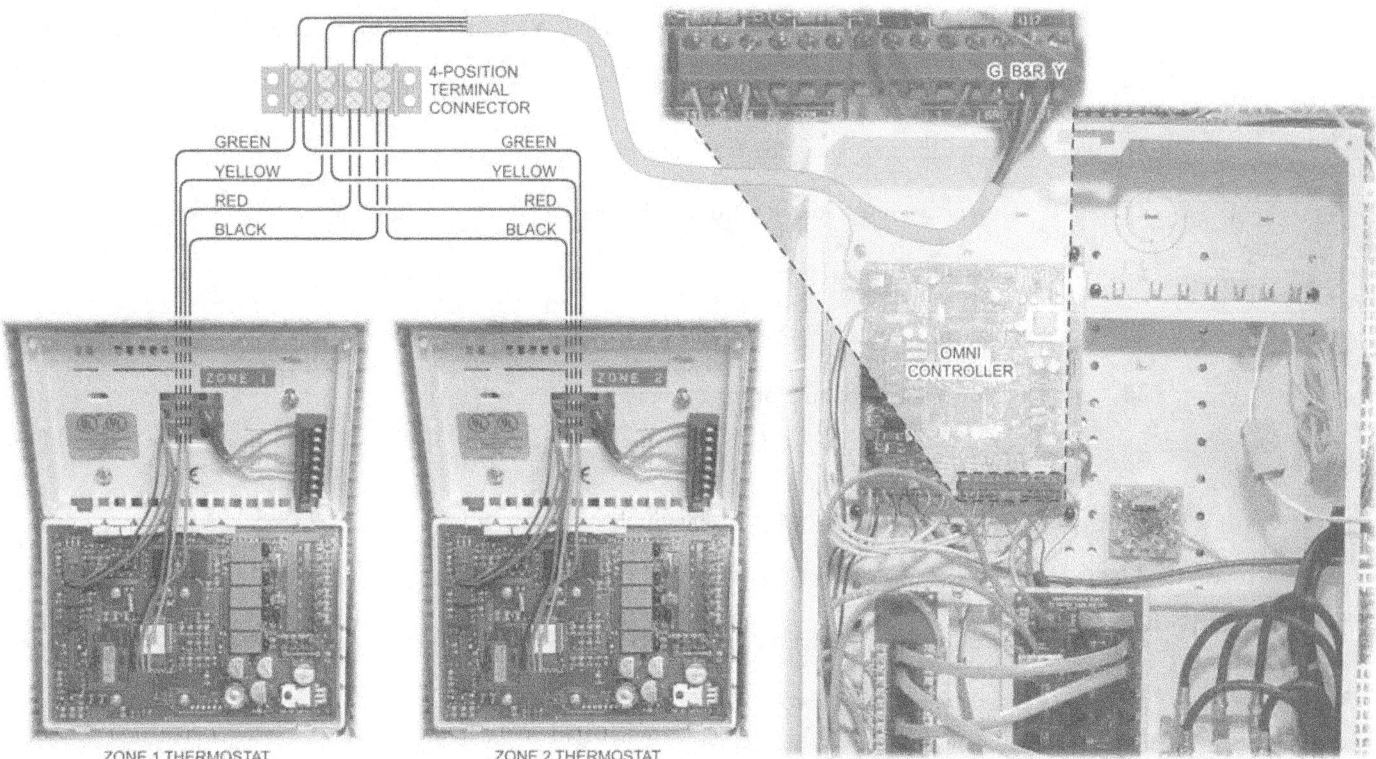

Figure 14-1: Communicating Thermostat Connections to the Controller

4. Remove the downstairs thermostat face from its mounting plate on the HVAC panel.

5. Connect the four-wire connector to the **COMM** pins on the communicating thermostat.

6. Route the four colored wires to the back of the HVAC panel.

7. Carefully place the thermostat back onto the mounting plate.

8. Remove the upstairs thermostat face from its mounting plate on the HVAC panel.

9. Connect the four-wire connector to the **COMM** pins on the communicating thermostat.

10. Route the four colored wires to the back of the HVAC panel.

11. Carefully place the thermostat back onto the mounting plate.

12. Connect all the black and red wire leads from the thermostats and the black and red wire leads for the signal cable to the terminal strip.

13. Connect all the yellow wire leads from the thermostats and the yellow wire lead for the signal cable to the terminal strip.

14. Connect all the green wire leads from the thermostats and the green wire lead for the signal cable to the terminal strip.

15. Route the signal cable back to and into the controller housing.

16. Connect the green wire lead to the terminal labeled **GRN** in the *TSTAT* block on the controller.

17. Connect the red and black wire lead to the terminal labeled **BLK** in the *TSTAT* block on the controller.

18. Connect the yellow wire lead to the terminal labeled **YEL** in the *TSTAT* block on the controller.

19. Power up the HVAC system.

20. Put both of the thermostats in the **OFF** mode.

21. Press **Setup** on the integration controller's keypad.

22. Enter the installer code (1111).

23. Enter the Installer mode by pressing the # key.

24. Press the **down arrow** button for further options.

25. Press option **5** for *TEMP*.

26. Choose the temperature format of **Fahrenheit**.

27. Press **1** for Fahrenheit, Press # to save.

28. Press the **down arrow** button for *THERMOSTAT 1 TYPE*.

29. Press # to change, Press **1** for *AUTO HEAT/COOL*.

30. Press # to save changes.

31. Press the **down arrow** button for *THERMOSTAT 2 TYPE*.

32. Press # to change, Press **1** for *AUTO HEAT/COOL*.

33. Press # to save changes.

34. Press **Cancel** button three times to exit.

Naming the Thermostats

1. Press **Setup** on the integration controller's keypad.

2. Enter the installer code (1111).

3. Press the **down arrow** button for further options.

4. Select option **7** for the *NAME* menu.

5. Press the **down arrow** button for further options.

6. Select option **5** for the *TEMP* menu.

7. Refer to the Appendix B Table of the Users Manual for the character codes used to name the thermostats.

8. Name the *TSTAT 1:* thermostat **DOWNSTAIRS** by entering the values (**36, 47, 55, 46, 51, 52, 33, 41, 50, 51**).

9. Press the # button to store the name of the thermostat in the controller.

10. Press the **down arrow button** to change to the **TSTAT 2:** thermostat.

11. Name the *TSTAT 2:* thermostat **UPSTAIRS** by entering the correct numeric values.

12. Press the # button to store the name of the thermostat in the controller.

13. Press **Cancel** button three times to exit.

Testing the Thermostat's Mode Change

1. Press the **6** button for *STATUS*.

2. Press the **down arrow** button for further options.

3. Press option **5** for the thermostat reading for the *downstairs Zone 1* thermostat.

4. Press the **down arrow** button twice to display the upstairs thermostat reading for the **Zone 2** thermostat.

5. Press the **CANCEL** button twice to return to the main menu.

6. Press the **TEMP** button.

7. Press the # button to select the *downstairs* zone.

8. Press option **1** for *MODE*.

9. Press the **down arrow** button for further options.

10. Press option **3** for *AUTO*.

11. Press the **TEMP** button.

12. Press the **down arrow** button to display the *upstairs* zone for selection.

13. Press the # button to select the *upstairs* zone.

14. Press option **1** for *MODE*.

15. Press the **down arrow** button for further options.

16. Press option **3** for *AUTO*.

17. Verify that both of the HVAC thermostat's LCDs display **AUTO**.

18. Record the current temperature readings for the HVAC thermostats in Table 14-1.

Table 14-1: Room Temperature

Current Room Temperature	
Current Room Temperature for Zone 1	
Current Room Temperature for Zone 2	

Testing the Downstairs and Upstairs Zones for Heating and Cooling

1. Press the **TEMP** button.

2. Press the # button to select the *downstairs* zone.

3. Choose option **2** for HEAT.

4. Enter a temperature into the integration controller's keypad that is five degrees warmer than the current reading for the downstairs thermostat recorded in Table 14-1.

5. Press the # button to store the value in the integration controller.

6. Wait for the HVAC unit to call for the heat cycle.

7. Verify that the remote damper for the *upstairs* zone is **closed**.

8. Wait for the HVAC system to finish the heating cycle.

9. Use the integration controller's keypad to decrease the heating value for the downstairs thermostat back to the current room temperature recorded in Table 14-1.

10. Press the # button to store the value in the integration controller.

11. Press the **TEMP** button.

12. Press the # button to select the *downstairs* zone.

13. Choose option **3** for *COOL*.

14. Enter the current temperature value recorded in Table 14-1 for the *downstairs* zone.

15. Press the # button to store the value in the integration controller.

16. Verify that the remote damper for the *upstairs* zone is **closed**.

17. Repeat steps 1 through 16 for the *upstairs* zone.

18. Verify the remote damper for the *downstairs* zone closes during the heating and cooling cycles.

19. Press the **TEMP** button.

20. Press the # button to select the *downstairs* zone.

21. Press option **1** for *MODE*.

22. Press option **0** for *OFF*.

23. Press the **TEMP** button.

24. Press the **down arrow** button to display the *upstairs* zone selection.

25. Press the # button to select the *upstairs* zone.

26. Press option **1** for *MODE*.

27. Press option **0** for *OFF*.

28. Turn **off** the HVAC system power.

LAB QUESTIONS

1. What are the two major types of thermostats?

2. What is another name for a intelligent thermostat?

3. What are some of the capabilities of an intelligent thermostat?

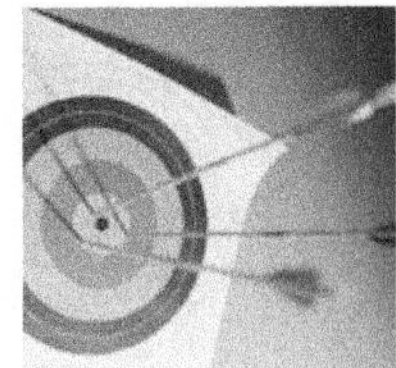

Feedback

Operating the HVAC Panel

OBJECTIVES

1. Run the HAI Web-Link II software.
2. Log in to the Web-Link II home page.
3. Perform various commands on the equipment using the computer software.
4. Log out of the HAI Web-Link II software.
5. Program various HVAC events through the integration controller keypad.
6. Use a touch tone phone to program HVAC zones.

HVAC Systems

RESOURCES

1. Marcraft HVAC Experiment Panel and Frame
2. Marcraft Integration Panel and Frame
3. Computer system with Windows XP Professional, IIS, and HAI Web-Link II software installed

DISCUSSION

Various manufacturers have developed access systems that enable the homeowner to control the residential HVAC system remotely through telephone operations, X-10 controllers, and Internet connections. Truly convenient HVAC control can be achieved by including a remote access system that will enable the owner to make changes to the HVAC system even when away from the residence.

A typical remote access design can employ various methods to establish a remote connection. In this procedure a connection will be established through a computer server connected to the HVAC equipment. The user interface for the connection will be a computer software browser window designed for use with the installed system. The computer will require the user to log in to the system for identity verification. Once the user's identity is verified, the user is granted access to the system. System components that can be commanded will appear in the browser window.

A typical scenario is the homeowner changing temperature settings at the residence while away from home. The homeowner may want to raise or lower the temperature a few degrees before he or she comes home. Or they may simply want to check the status of the system from work during adverse weather conditions. The user connects to the residence's server via a browser interface and makes the desired changes to the system.

PROCEDURES

HVAC Systems

HVAC Zones Programmed Through the Web Client

1. Start the computer system.

2. Power up the Marcraft HVAC Experiment Panel.

3. Verify that the wire connections between the Marcraft HVAC Experiment Panel and the Integration Controller Panel are still connected.

4. Start the *Web-Link Server* program, if it was not started during the computer boot sequence.

NOTE: If running, the Web-Link Server program will appear as a house icon in the taskbar.

5. Click the **Web-Link Web Client** icon on the desktop.

6. Enter the **User Name** _______ and the **User Code** _______ in the *HAI Web-Link II* login screen. See Figure 15-1.

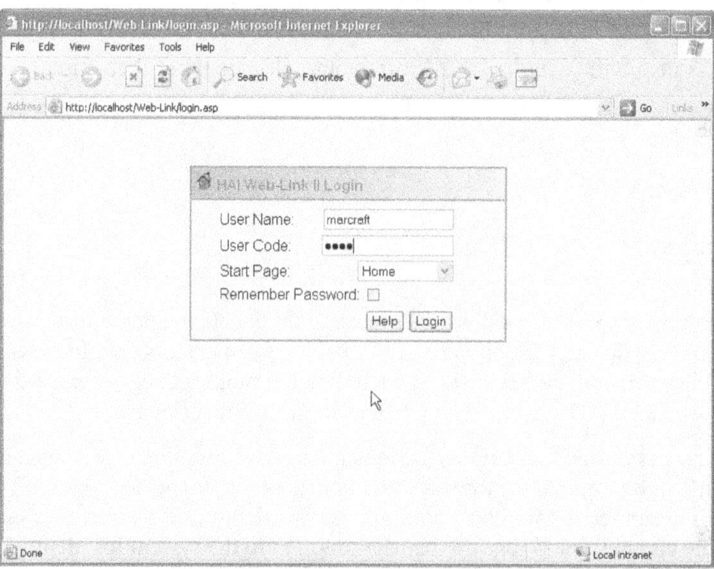

**Figure 15-1:
HAI Web-Link II
Login Screen**

NOTE: Enter this information correctly. Only three attempts are allowed. If three unsuccessful attempts are made the system will issue a one-hour lockout.

7. Click the **Login** button.

8. Click the **Temperature** button. See Figure 15-2.

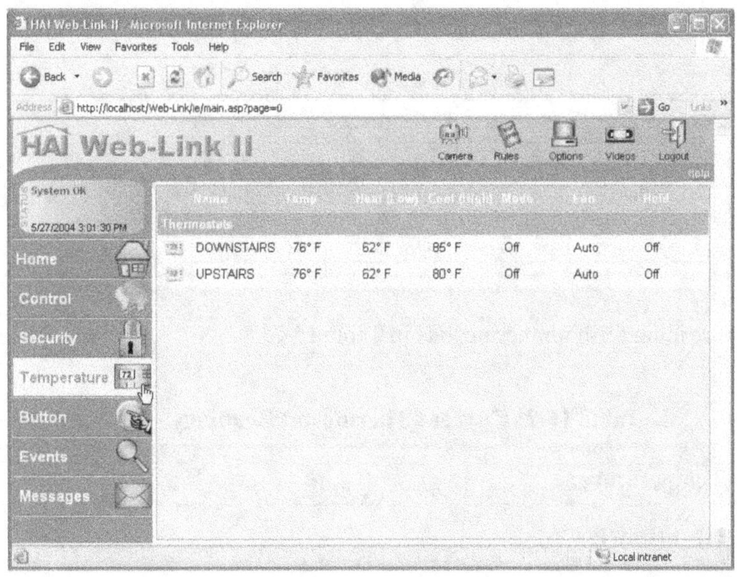

Figure 15-2:
Temperature Button
Window

9. In the right window pane, click the word **DOWNSTAIRS**.

10. Click the **down arrow** button next to the word *Command* to display available options. See Figure
15-3.

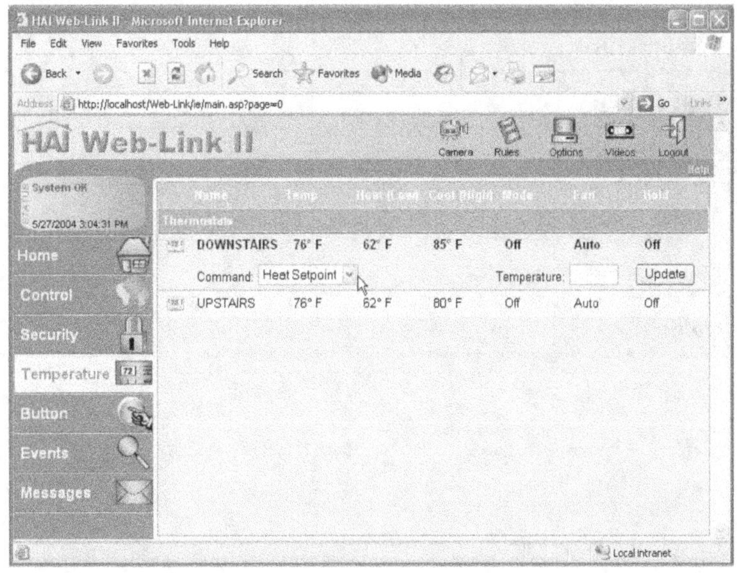

Figure 15-3:
Seeking Command
Options for
Downstairs

11. Record the available options in Table 15-1.

Table 15-1: Command Options for Temperature

12. Record the current reading for both thermostats in Table 15-2.

Table 15-2: Current Thermostat Readings

Downstairs Thermostat	**Temperature:**	
	Heat (Low):	
	Cool (High):	
	Mode:	
	Fan:	
	Hold:	
Upstairs Thermostat	**Temperature:**	
	Heat (Low):	
	Cool (High):	
	Mode:	
	Fan:	
	Hold:	

13. Change the *Mode* for both thermostats to **Heat**. See Figure 15-4.

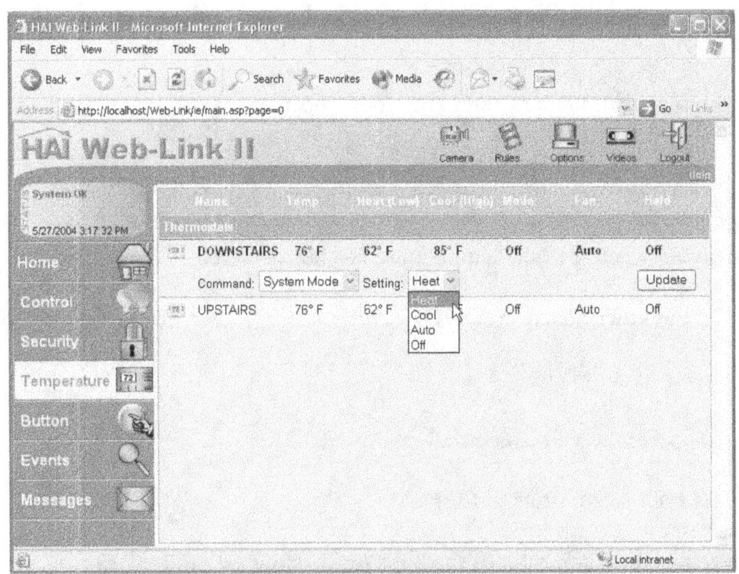

**Figure 15-4:
Changing Mode to
Heat**

14. Change the *Heat Setpoint* for the *Downstairs* thermostat **up five degrees**, as shown in Figure 15-5.

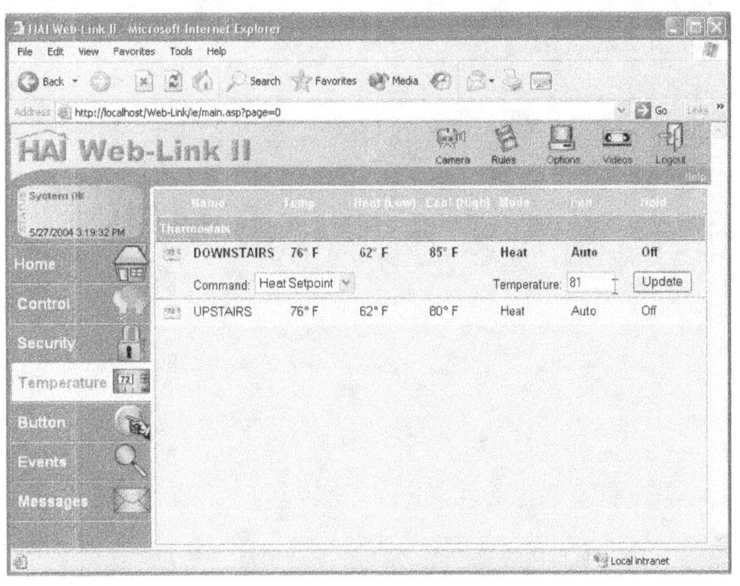

**Figure 15-5:
Setting the Heat
Setpoint Up Five
Degrees**

15. Click the **Update** button when the new temperature has been entered.

16. Change the *Heat Setpoint* for the *Upstairs* thermostat **up five degrees**.

17. Click the **Update** button when the new temperature has been entered.

18. Allow the heating cycle to complete.

19. Set the *Heat Setpoint* back to the setting recorded in Table 15-2.

20. Change the *Cool Setpoint* for the *Downstairs* thermostat **down five degrees** lower than the new current temperature.

21. Click the **Update** button when the new temperature has been entered.

22. Change the *Cool Setpoint* for the *Upstairs* thermostat **down five degrees** lower than the new current temperature.

23. Click the **Update** button when the new temperature has been entered.

24. Change the *Mode* on both thermostats to **COOL**.

25. Allow the cooling cycle to complete.

26. Return the *Cool Setpoint* to the value recorded in Table 15-2.

27. Change the *Mode* for both thermostats to **OFF**.

28. Click the **Logout** button at the top right of the browser window. See Figure 15-6.

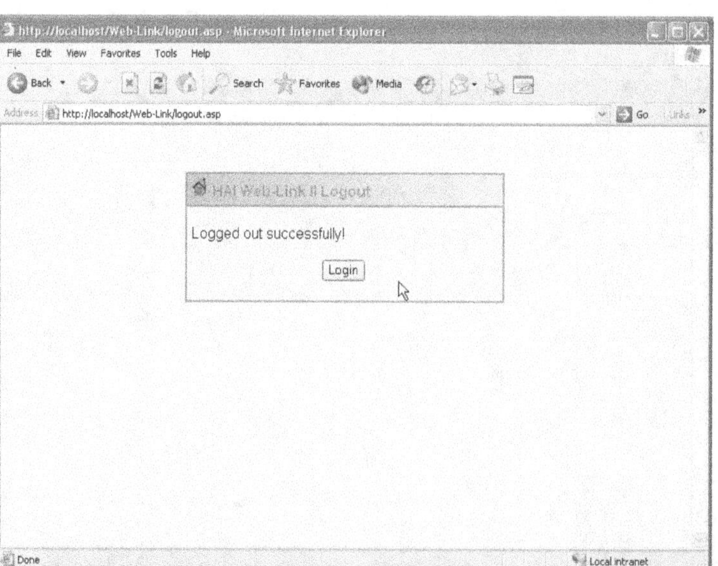

**Figure 15-6:
Logged Out
Successfully**

HVAC Zones Programmed Through the Integration Controller Keypad

Programming the Timing Events (Morning Warming)

1. Press the **SETUP** button.

2. Enter the *Installer Code* (**1111**).

3. Press **3** for *PROGRAM*.

4. Press **1** for *ADD*.

5. Press **2** for *CMD*.

6. Press the **down arrow** button for further options.

7. Press **5** for *TEMP*.

    Menu Display: DOWNSTAIRS
    ENTER TEMP ZONE 0=ALL

8. Press **0** for all and # to continue.

    Menu Display: ALL TSTATS
    1=MODE 2=HEAT 3=COOL ↓
    4=FAN 5=HOLD

9. Press **2** for *HEAT*.

10. Enter **75** for the temperature and the # button to store the value.

11. Press **1** for *WHEN*.

12. Press **1** for *TIMED*.

13. Press **1** for *TIME*.

14. Enter **700** and press the **up arrow** button.

    Menu Display: 7:00AM 5/21
    1=TIME 2=DATE/DAY

NOTE: To run this program during the class period, the start time for the event must be coordinated with the current time and room temperatures. Adjust the start time value to the current class time and add a few minutes extra for programming requirements. Set the heating temperature 5 degrees above the current room temperature so that a heating call will be made.

15. Press **2** for *DATE/DAY*.

    Menu Display: DATE: 5/21
                         MMDD ↓=DAY

16. Press the **down arrow** button to select *Day*.

    Menu Display: DAY (S): ONCE
                         1-7=MON-SUN 0=ONCE

17. Press the following buttons: **1, 2, 3, 4**, and **5**.

    Menu Display: DAY (S): MTWTF__
                         1-7=MON-SUN 0=ONCE

18. Press the # button four times. The program will be stored when the controller issues a single beep.

19. Press the **CANCEL** button twice to return to the main menu.

Programming the Timing Events (Evening Cooling, Warming Set Back)

1. Press the **SETUP** button.

2. Enter the *Installer Code* (**1111**).

3. Press **3** for **PROGRAM**.

4. Press **1** for *ADD*.

5. Press **2** for *CMD*.

6. Press the **down arrow** button for further options.

7. Press **5** for *TEMP*.

    Menu Display: DOWNSTAIRS
                         ENTER TEMP ZONE 0=ALL

8. Press **0** for all and # to continue.

    Menu Display: ALL TSTATS
                         1=MODE 2=HEAT 3=COOL ↓
                         4=FAN 5=HOLD

9. Press **2** for *HEAT*.

10. Enter **65** for the temperature and the # button to store the value.

11. Press **1** for *WHEN*.

12. Press **1** for *TIMED*.

13. Press **1** for *TIME*.

14. Enter **500** and press the **down arrow** button.

Menu Display:　5:00PM 5/21
　　　　　　　　1=TIME 2=DATE/DAY

NOTE: To run this program during the class period, the start time for the event must be coordinated with the current time and room temperatures. Adjust the start time value to the current class time and add a few minutes extra for programming requirements. Set the heating temperature 5 degrees above the current room temperature so that a heating call will be made.

15. Press **2** for *DATE/DAY*.

Menu Display:　DATE: 5/21
　　　　　　　　MMDD ↓=DAY

16. Press the **down arrow** button to select *Day*.

Menu Display:　DAY (S): ONCE
　　　　　　　　1-7=MON-SUN 0=ONCE

17. Press the following buttons: **1, 2, 3, 4,** and **5.**

Menu Display:　DAY (S): MTWTF__
　　　　　　　　1-7=MON-SUN 0=ONCE

18. Press the # button four times. The program will be stored when the controller issues a single beep.

19. Press the **CANCEL** button twice to return to the main menu.

Checking the HVAC Programmed Events

1. Press the **SETUP** button.

2. Enter the *Installer Code* (**1111**).

3. Press **3** for *PROGRAM*.

4. Press **2** for *SHOW*.

5. Press # for *EVERY*.

6. Press the **down arrow** button until the following menu is displayed.

Menu Display: 7:00AM MTWTF—
                  ALL TSTATS HEAT 75

7. Press the **down arrow** button again until the following menu is displayed.

Menu Display: 5:00 PM MTWTF—
                  ALL TSTATS HEAT 65

8. Press the **CANCEL** button until the main menu is displayed.

HVAC Zones Programmed Through the Touch-Tone Phone

SETUP NOTE: Connect the integration controller's telco line (white/blue, blue) to the CO LINE 8 phone line simulator's right port. Connect the Remote Location 1 phone to the CO LINE 8 phone line simulator's left port. See Figure 15-7.

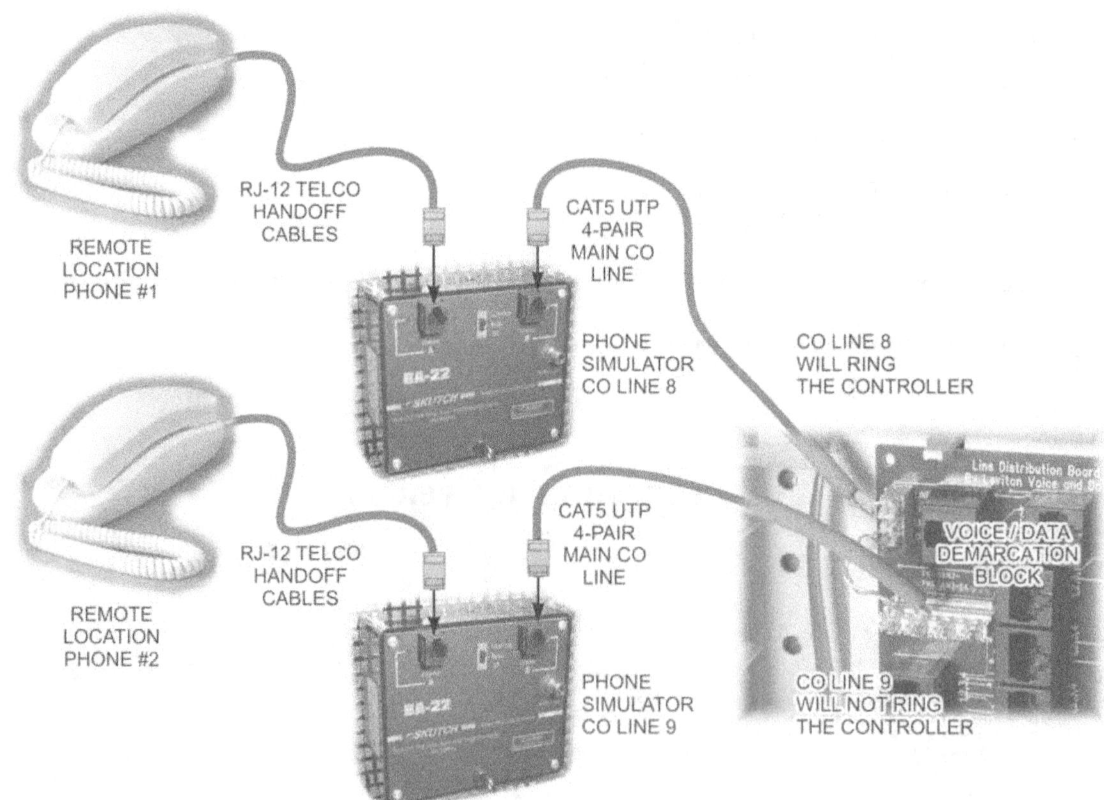

Figure 15-7:
Touch-Tone Phone
and Controller
Connection Setup

Entering the Main Menu

1. Pick up the Remote Location **1** phone.

2. Let the phone ring until the controller beeps.

3. Enter the master code ____ within the first 3 seconds.

4. The voice message"Welcome to Omni. Please choose..." will be heard.

5. Listen to the following menu selections:

     1 Control
     2 Security
     3 Button
     4 All
     5 Temperature
     6 Status
     7 Events
     8 Phone
     9 Goodbye
     * Cancel
     0 Repeat

Entering Into Submenus

1. Press **5** for *Temperature*.

2. The voice message "Temperature. Enter temperature number then #" will be heard.

3. Enter **1#** for the *Zone 1* thermostat.

4. The voice message "Thermostat 1 temperature is 76 degrees. Please choose..." will be heard.

5. Listen to the following menu selections:

     1 Mode
     2 Heat Setting
     3 Cool Setting
     4 Fan
     5 Hold
     # Status
     * Cancel

6. Choose option # for *Status*.

7. The voice message "Temperature is 76 degrees. Heat setting is 76 degrees. Cool setting is 79 degrees. Mode is off. Fan is off. Hold is off. Please choose..." will be heard.

8. Record the values from the voice message in Table 15-3.

Table 15-3: Current Class Room Thermostat Variables

Zone 1	
Temperature is:	
Heat Setting is:	
Cool Setting is:	
Mode is:	
Fan is:	
Hold is:	
Zone 2	
Temperature is:	
Heat Setting is:	
Cool Setting is:	
Mode is:	
Fan is:	
Hold is:	

NOTE: The variables in the message will differ depending upon the current class room temperatures and the state in which the unit was left.

9. Listen to the following menu selections:

    1 Mode
    2 Heat Setting
    3 Cool Setting
    4 Fan
    5 Hold
    # Status
    * Cancel

10. Check the status for the Zone **2** thermostat.

11. Record the variables for the Zone 2 thermostat in Table 15-3.

12. Select option **1** for *Mode*.

13. Listen to the following menu selections:

        0 Off
        1 Heat
        2 Cool
        3 Auto
        * Cancel

14. Select option **3** for *Auto*.

NOTE: The controller will revert back to the main menu.

15. Navigate back to the *temperature* menu.

16. Change the *Heat Setting* value for both thermostats *down five degrees* from the current value recorded in Table 15-3 by selecting the **Heat Setting** option.

17. Change the *Cool Setting* value for both thermostats *down five degrees* from the current value recorded in Table 15-3 by selecting the **Cool Setting** option.

18. Check the status of both thermostats and record the variables in Table 15-4.

Table 15-4: Adjusted Thermostat Variables

Zone 1	
Temperature is:	
Heat Setting is:	
Cool Setting is:	
Mode is:	
Fan is:	
Hold is:	
Zone 2	
Temperature is:	
Heat Setting is:	
Cool Setting is:	
Mode is:	
Fan is:	
Hold is:	

NOTE: The variables in the message will differ depending upon the current class room temperatures and the state in which the unit was left.

19. Use the touch-tone phone to return the thermostats to the **OFF** mode.

20. Use the Omni console to enter the set up programs menu, and press the **3** (*DELETE*) key. The console displays:

     DELETE ALL PROGRAMS?
     0=NO 1=YES

21. To permanently delete all the automation programs in the system, press the **1** (*YES*) key.

22. Turn the Marcraft HVAC Experiment Panel **OFF**.

23. Perform a general cleanup of your lab area, and return all applicable equipment and components to their proper storage locations.

Environmental Control Objective Map

There are two levels of expertise proposed for those workers who install electronics cables in residences and interconnect electronics communications, computer, control or entertainment equipment. **RESI, the Residential Electronics Systems Integrator**; and the **Master RESI, Residential Electronics Systems Integrator.**

The RESI integrator is proficient in pre-wiring for home theater and telecommunications equipment interconnection. He/she will install wiring for antenna, satellite, cable TV, wireless broadband, audio/video entertainment and computer equipment/networking. He/she will have skills required for low voltage wiring. Workers who hold the RESI certification will likely also hold one or more of the 5 specialty endorsements depending on the area of expertise that worker requires.

The MASTER RESI will be proficient at all of the RESI skills and knowledge as well as in planning and designing electronics and communications equipment systems and layout for new and existing construction. The RESI worker is capable of designing the entire system and network for audio, video, data and control of security and environment. He/she also is capable of troubleshooting and debugging the system and planning installation or modifications. The MASTER RESI has extensive knowledge of the operation and technology and is proficient in each of the basic five subcategories of residential electronics.

Domain 1.0   Lighting

1.1 Describe basic lighting controls, on/off, dimmer and X10

1.2 Compare common remote control methods and list advantages or disadvantages of each

1.3 Explain time or event automatic control operation

1.4 Describe how lighting can be programmed for multiroom and zones control

Domain 2.0   Computer Interfacing

2.1 Describe a home local area computer network - LAN

2.2 Compare bus, Star and Ring network configuration

2.3 List hardware components needed for home network systems

2.4 Compare available home automation software

2.5 List usage of wireless control of elements of the home automation/control system

2.6 Describe the role the RESI may assume in implementing the completed control installation and programming/configuration

Domain 3.0   Sensors & Actuators

3.1 Describe sensors: (Rain, Temperature, Motion, etc.) and show where and how they are used

3.2 Describe actuators and relays and their applications

Domain 4.0   Computer Control

4.1 Name commercially available computer control programs for home control and automation

4.2 Describe the steps required to program the computer to control the home system

Domain 5.0   Needs Assessment

5.1 Explain the considerations involved in planning the control system and in tying it in with the other home technologies, with documentation and compliance with local or state codes

5.2 Explain the steps in establishing a timetable for customer approval, product procurement and installation sequences

Domain 6.0   HVAC Interfacing

6.1 Describe the advantages of interfacing the network with the heating, ventilation and cooling system of the home

6.2 Describe the options a control system may offer for programming thermostat control of living area HVAC processes

Domain 7.0   Event Recording - Storage

7.1 Describe why and how home events data may be accessed

7.2 Explain how event sequences can be incorporated into a home control system and the advantages of doing so

Domain 8.0   Home Theater

8.1 Outline the advantages of including annunciating and surveillance equipment on-screen in the home theater-viewing areas

8.2 Diagram how mechanized equipment may be activated or automated from the home theater viewing area or command center

8.3 Discuss the extent to which maximum remote control functions can be integrated into a master control remote control hand unit with secondary RHUs also active in other locations in the home

Domain 9.0   System Control

9.1 Describe how programmable logic control (PLC) is utilized in home control systems

9.2 Describe the features of Macro Designer

9.3 Describe the X10 technology and how it may be the most practical control method for some applications

9.4 Describe the HAI Omni Controller

Domain 10.0   Implementation

10.1 Explain how lawn water sprinkler systems can be incorporated in the home control system, mechanically connected and programmed

10.2 Explain hoe yard or in-home water fountains are interfaced and controlled by the home automation system

10.3 Describe how spas and hot tubs may need different control algorithms than other home features

10.4 Describe how the entire system can be controlled and accessed via wireless or phone connections from remote locations

Glossary

10Base-T: An Ethernet specification defined by the IEEE 802.3 committee officially as specification 802.3i. The specification describes a 10 Mbps wire speed using baseband information. The specification requires a star topology using Category 3, 4, or 5 unshielded twisted pair (UTP) cable.

100Base-T: An Ethernet IEEE 802.3u specification with a wire speed of 100 Mbps using Category 5 or 5e UTP cable.

1000Base-T: An IEEE 802.3ab standard for Gigabit Ethernet. Gigabit Ethernet runs at 1,000 megabits per second using Category 5e cable.

A

AC cable: Armored cable is identified by the code name AC cable. AC cable is often referred to as *BX*, although this is the trademark of a specific manufacturer. Armor clad and metal clad (MC) might sound like the same cable, but some important differences exist. AC cable uses the interior bond wire, in combination with the exterior interlocked metal armor, as the equipment grounding means of the cable. MC cable, on the other hand, is manufactured with a green insulated grounding conductor, and this conductor, in combination with the metallic armor, comprises the equipment ground. AC cable can have up to four insulated conductors only; a fifth insulated conductor is allowed by UL, if it is a grounding conductor. Each conductor in AC cable is paper-wrapped.

AC-3 (Audio Coding-3): Dolby's digital audio data compression algorithm adopted for HDTV transmission, laserdiscs, and CDs for 5.1 surround sound multichannel home theater use.

ADSL terminal unit (ATU): The engineering term used for describing an ADSL modem. The ADSL terminal unit is also known as the ADSL transceiver unit, ADSL transmission unit, and ADSL termination unit. The ATU-R (remote) unit is installed in the home, and the ATU-C (central) office unit is installed in the local telephone company central office.

Advanced Television Systems Committee (ATSC): The group that developed voluntary national standards for high-definition television (HDTV), standard-definition television (SDTV), data broadcasting, multichannel surround sound audio, and satellite direct-to-home broadcasting.

air handler: The subassembly contained in an HVAC air distribution system that typically houses items such as the supply fan(s), return fan(s), heating coil, cooling coil, and filters. Its purpose is to provide the pressure necessary to get the airflow through the entire system, and at the desired temperature and humidity.

alarm system: A dedicated electronic system using a central controller and distributed sensors that detects unauthorized intrusion into a home interior, or protected external areas. It is typically used with door and window sensors, glass break sensors, motion detectors, and occasionally closed circuit television. When armed by the user, the system detects intrusion from the outside and can react with responses that vary from the activation of interior sirens, bells, and lighting, to the notification of a central alarm monitoring system.

American National Standards Institute (ANSI): A private, nonprofit, nongovernmental national organization that serves as the primary coordinator of standards within the United States, in relation to ISO standards.

American Wire Gauge (AWG): Standard measuring gauge for nonferrous conductors. Gauge is a measure of the diameter of the conductor. The higher the AWG number, the thinner the wire.

ampere: A measuring unit for the rate of electron flow or current in an electrical conductor. One ampere of current represents one coulomb of electrical charge (6.24×10^{18} charge carriers) moving past a specific point in one second. The ampere is named after Andre Marie Ampere, a French physicist (1775-1836).

amplifier bridging: A technique used to configure a two-channel stereo amplifier to drive a single load (speaker) with more power than the sum of the two original channels. An amplifier running in bridged mode has a single output channel to which a load (speaker) can be connected.

amplitude: The magnitude of a signal in voltage or current. It's frequently expressed in terms of peak, peak-to-peak, or RMS.

analog: A term used in telecommunications technology to describe a continuously variable waveform that is analogous to the electrical signal produced by the human voice when processed by a microphone and audio amplifier. Also, it's a signal in which a base carrier's alternating current frequency is modified in some way, such as by amplifying the strength of the signal, or varying the frequency to add information to the signal (also called *amplitude modulation*). Broadcast and telephone transmissions have conventionally used analog technology. An analog signal can be represented as a series of sine waves. The term originated because the modulation of the carrier wave is analogous to the fluctuations of the human voice or other sound that is being transmitted.

anti-siphon valve: A control valve with a built-in atmospheric vacuum breaker (backflow preventer). Most commonly used in residential irrigation systems.

apparent power: In an AC circuit apparent power is the assumed value of P = E (volts) and I (amperes) and is expressed as volt-amperes (VA). Apparent power is the theoretical power consumed in a circuit, while neglecting the power factor. (See "true power.")

asymmetric digital subscriber line (ADSL): Asymmetric DSL is a class of DSL service described in ANSI TI.413-1998 that offers higher download speeds with a slower upstream speed. The actual network bandwidth a customer receives from a DSL service in the home depends on the total length of the telephone local loop distance to the local office.

Atmospheric Vacuum Breaker (AVB): A type of backflow preventer. (See "valves.")

Audio Engineering Society (AES): Founded in 1949, the largest professional organization for electronic engineers and all others actively involved in audio engineering. It's primarily concerned with education and standardization.

audio/video (A/V) receiver: A component of a home audio/video system. A/V receivers have an integrated AM/FM tuner, and process from 4 to 6 channels with amplifiers capable of driving multiple speaker channels. They have built-in surround sound decoders that separate the sound signal from a video source (like a DVD player) into several channels, then route it to several speakers.

B

backflow prevention valve: A valve used with irrigation systems that prevents potable water from being contaminated with nonpotable water in the event a cross-connection exists during a condition of backflow.

backwash valve: A valve that reverses the flow of water in a swimming pool filter during the filter cleaning cycle. Modern backwash valves are molded of temperature- and chemical-resistant chlorinated polyvinyl chloride (CPVC) material.

bandwidth: The transmission capacity of a medium expressed as a range of frequencies in hertz (Hz). A greater bandwidth indicates the capability to transmit a greater amount of data over a given period of time.

Bayonet Neil Concelman (BNC) connector: A coaxial cable connector that, in its male form, has a center pin (bayonet) connected to the center conductor and a metal tube connected to the cable shield. A rotating ring surrounds the tube and contains small holes that mate with projections on the female connector to make the connection. The connector was developed in the late 1940s and is named after the creators — Amphenol engineer Carl Concelman and Bell Labs engineer Paul Neill. Neill designed the N-type connector, and Concelman designed the C-type connector. The BNC is a hybrid N/C-type with a mechanical extra appropriately called a *bayonet*. BNC connectors are often identified using other terms of unknown origin, such as British naval connectors, bayonet navy connectors, or bayonet nut connectors.

bimetal element: A type of temperature sensor used in older thermostats containing two strips of different metals that are attached together. As the room temperature changes, the lengths of the metal strips change. Because the metals are different and change their lengths at different rates, they are used to operate thermostat switches that react to temperature changes.

bit (binary digit): The smallest component of information in a binary notation system. A bit is a single 1 or 0.

bit rate: The number of binary digits transmitted over a medium per unit of time. Bit rate is specified as the number of bits transmitted in one second (bps).

Bluetooth: A wireless technology standard designed to connect one device to another with a short-range radio link. This standard has evolved from early work and engineering studies performed by Ericsson Mobile Communications in 1994. The IEEE Project 802.15.1 has derived a Wireless Personal Area Network (WPAN) standard based on the Bluetooth v1.1 Foundation specifications.

booster pump: A device that increases the water pressure in a system where some pressure already exists.

branch circuits: The circuits in a house that branch from the service panel to boxes and devices.

breaker: A switch-like device that opens a circuit if the current to the branch circuit exceeds the specified rating of the device.

Broadband over Power Line (BPL): A system that proposes to use the electric power distribution grid for delivering high-speed connections to the Internet. On April 23, 2003 the FCC announced a Notice of Inquiry seeking public comment on what it calls "Broadband over Power Line" (BPL): the use of existing electrical power lines as a transmission medium to provide high-speed communications capabilities, including Internet and broadband services, to both urban and rural areas by coupling radio-frequency energy onto the power line.

Btu (British thermal unit): The amount of thermal energy necessary to raise the temperature of one pound of water by one degree Fahrenheit at sea level.

bundled cabling: A type of cable available from vendors described in addendum 3 to the ANSI/TIA/EIA 568A standard that makes installation of Grade 2 structured wiring easier. The addendum describes four-pair cable assemblies that are not covered by an overall sheath (as specified for hybrid cables) but by any binding method, such as speed-wrap or cable ties. A standard bundled cable for Grade 2 consists of two Category 5e cables and two RG-6 quad-shield cables.

byte: A group of bits generally accepted in the computer industry as 8 bits (but 9 bits on 36-bit machines). Some older computer architectures used "byte" for quantities of 6 or 7 bits but these usages are now obsolete. An octet always contains 8 bits, and this term is used to avoid confusion with the term "byte."

C

cable tester: A type of network cable test equipment used to check cables for opens, shorts, and continuity.

call waiting: A local exchange telephone service option that notifies a subscriber conducting a call that one or more other phones are trying to make a connection. With call waiting service, the second call is announced by a soft beep. The user can ask the first caller to wait while she answers the second call. She then has the option to return to the first call or disconnect from the first to talk to the second caller.

caller ID: A local exchange telephone system service that identifies the number of the caller to the receiving (called) party. This service is one of many listed under the title of caller line identification or calling line identity (CLI) services. Features such as caller display and call return use a feature of modern telephone networks that transmits the number of the caller as each call is set up. Called customers using these CLI services have access to this information. With caller display, a customer can see the number on his phone before answering the call.

Category 5: A type of UTP communications cable described in ANSI/TIA/EIA 568A that has four twisted pairs of wires. Category 5 cable is tested to 100 MHz. Category 5 cable was previously used in residential structured wiring and 100Base-T Ethernet LAN installations. It has been replaced by Category 5e as the preferred structured wiring cable for new installations, which provides improved performance at approximately the same cost.

Category 5e: An enhanced version of Category 5 cable described in specification ANSI/TIA/EIA 568A addendum 5. Category 5e cable was officially ratified as a standard on December 13, 1999, and updated with the release of ANSI/TIA/EIA-568B.1-2000. Although it has a rated bandwidth of 100 MHz, it has improved specifications for near-end crosstalk (NEXT), power-sum equal-level far-end crosstalk (PSELFEXT), and attenuation. The specification also enforces several attributes that were optional in the original Category 5 specification. Category 5e cable is usually tested to a bandwidth of 350 MHz, despite its 100 MHz specified bandwidth.

Category 6: A type of data communications cable that complies with the transmission requirements in the TIA/EIA 568B.2-1 Commercial Building Telecommunications Category Standard. Cable carrying the Category 6 rating has a rated bandwidth of 250 MHz.

Category 7: A type of high-performance data communications twisted-pair cable that uses a braided shield surrounding all four foil-shielded pairs to reduce noise and interference. Category 7 cable, connecting hardware, and patch cords are rated to a maximum frequency bandwidth of 600 MHz.

Cathode-Ray Tube (CRT): A type of vacuum tube in which images are produced when an electron beam strikes a phosphorescent surface. CRTs are used as the primary video display device in computer monitors and television receivers.

CCTV: Abbreviation for closed circuit television.

CEBus (Consumer Electronics Bus): CEBus is a communications protocol developed by the Electronic Industries Alliance (EIA) as the ANSI/EIA-600 standard. CEBus is designed to control devices on a power line, but it also works on other media. The CEBus standard involves device addresses that are set in hardware at the factory, and includes four billion possibilities.

cipher locks: A secure access control system for doors, using personal access codes known by the user. Cipher locks operate by unlocking a door equipped with magnetic locks, when the correct programmed code is entered by the user on the cipher lock keypad.

cladding: The portion of a fiber-optic cable encircling the core material.

composite video: A single video signal that contains luminance, color, and synchronization information. NTSC, PAL, and SECAM are all examples of composite video systems.

condenser: A heat exchanger (outdoor coil) assembly used with residential air conditioning or heat pump systems. As an integral component of a heat pump or a central air conditioning system, it is located in a vented outdoor casing where a large fan passes air over the coils to exhaust heat. It is the transfer point where heat is transferred from inside the home to the outside air.

conduit: A type of enclosure for wiring, installed in hazardous locations such as potentially explosive air environments, underground wiring, or outdoor and moist locations. It is available in a variety of types for many environments. They range from rigid metal conduit, to intermediate metal conduit, to thin-walled conduit called electrical metallic tubing (EMT), to nonmetallic types called polyvinyl chloride (PVC).

copper communications riser cable (CMR): A fire rating for riser cable established by the National Electrical Code (NEC) ANSI/NFPA 70 standard. The NEC assigns fire ratings to communications transmission cables and discusses suitable applications for each cable type. Copper communications riser cable is suitable for installation in vertical riser shafts that run from floor to floor. It cannot be used in any environmental air spaces, such as air conditioning ducts, unless the local code allows it. CMR cable might also be labeled as flame test (FT)-4, and is not as expensive as plenum cable.

cross-connect: A facility enabling the termination of cable elements and their interconnection, and/or cross-connection, primarily by means of a patch cord or jumper.

crossover cable: A cable used to directly connect two computers, or two hubs, in which the receive and transmit wires of a twisted-pair cable have been transposed at each end of the cable connector to perform the crossover function normally performed by a single hub.

crossover networks: A device used with multiple speaker systems to separate portions of the audio spectrum, and routing them to individual speakers for optimum performance.

D

daisy chain: A type of residential wiring design used in home lighting systems, in which load requirements for individual lighting systems do not require a dedicated home-run connection to the main power distribution panel. Some lighting systems include multiple light fixtures controlled with wall switches that share a 15-ampere daisy-chain service from the main power distribution panel.

digital: Circuitry or telecommunications transmission systems in which information-carrying signals are restricted to either of two states, corresponding to logic 1 or 0.

Digital Subscriber Line (DSL): A high-speed Internet access service offered by local telephone companies using regular telephone lines. DSL can move data over the phone lines at speeds up to 6 Mbps, or 140 times faster than the fastest analog modems (56,000 bits per second). In addition to its very high speed, DSL has many benefits over analog connections. Unlike dial-up connections that require analog modems to dial in to the Internet service provider every time, DSL connections are always on. Another benefit is that the user can talk on the telephone at the same time he is accessing the Internet.

Digital Versatile Disc (DVD): A high-volume storage media with the same physical appearance and dimensions as a compact disc. Initially called digital video disks by the creators, DVDs were named digital versatile disks in 1996. DVDs are standardized with five physical formats and four media storage versions.

DVD-ROM is a high-capacity data storage medium. DVD-Video is a digital storage medium for feature-length motion pictures, and DVD-Audio is an audio-only storage format similar to CD-Audio. DVD-R offers a write-once, read-many storage format akin to CD-R. DVD-RAM was the first rewritable (erasable) flavor of DVD to come to market and has subsequently found competition in the rival DVD-RW and DVD+RW formats. With the same overall size as a standard 120-mm diameter, 1.2-mm thick CD, DVD discs provide up to 17 GB of storage with transfer rates that are higher than those of CD-ROMs, and access times similar to CD-ROMs. DVDs come in four versions. DVD-5 is a single-sided, single-layered disc, boosting capacity sevenfold to 4.7 GB. DVD-9 is a single-sided, double-layered disc offering 8.5 GB, whereas DVD-10 is a 9.4 GB dual-sided, single-layered disc. Finally, DVD-18 increases capacity to a huge 17 GB on a dual-sided, dual-layered disc. DVDs are similar in appearance to CDs — they are both 4 ¾ inches in diameter. DVD tracks, however, are closer together and the compression ratio is much higher. The result is that DVD discs can hold 4.7 GB per side compared to approximately 660 MB for a CD.

DIN: A common type of connector used for video and computer interfaces. DIN is an acronym for *Deutsches Insitut für Normung eV*, a standards-setting organization for Germany.

Dolby Digital (formerly known as AC-3): A 5.1-channel surround sound format, and the standard for DVD sound tracks. Dolby Digital features five discrete Dolby Surround channels (front center, front left, front right, surround left, surround right; giving it the "5" designation) of full-frequency sound, plus a sixth channel for low-frequency effects (LFE).

Dual-Tone MultiFrequency (DTMF): A method of tone dialing using a 12-button keypad with dual tones representing each of the 12 buttons.

Dynamic Host Configuration Protocol (DHCP): A network protocol that enables individual computers on an IP network to obtain their IP address configurations from a DHCP server. This reduces the work necessary to administer a large IP network by an administrator. The DHCP server allocates an IP address to the PC from one of the scopes (the pools of addresses) it has available.

dynamic speaker: An audio reproduction device, also referred to as a cone speaker, in which the audio signal is applied to a voice coil that is part of the moving system.

E

ElectroMagnetic Interference (EMI): A type of disturbance occurring in communications circuits when unwanted electrical currents are induced into the wiring or components of an electrical system. EMI can also degrade the performance of wireless communications channels.

electromagnetic lock: A type of lock used to control access through a door where maximum security is required. It consists of an electromagnet mounted in the door frame and a matching plate installed in the door. The electromagnet can be energized from a switch by the person controlling the access. When energized, the electromagnet holds the plate with a strong magnetic field. The lock has no moving parts. The amount of pressure an electromagnetic lock can withstand is somewhat determined by its integrity rating. The Builders Hardware Manufacturers Association (BHMA) together with the American National Standards Institute (ANSI) has determined three integrity grades for electromagnetic locks. A Grade 3 integrity rating provides a locking strength of 500−900 lb. An integrity rating of Grade 2 means that the electromagnetic lock must be strong enough to withstand a pressure of 1,000−1,400 lb. And an integrity rating of Grade 1 means that the electromagnetic lock is designed to withstand pressures of 1,500−2,700 lb.

electromechanical controller: A type of irrigation system controller used to operate the valve stations at prescribed times. Electromechanical controllers use clocks in addition to stored irrigation programs. They are respected as a reliable solution for simple irrigation systems. They have few sophisticated electronic components and are driven by electric motors and gears. Turning dials or flipping switches programs the controller to select watering times, how long each zone will be watered, and on which days the watering will occur.

electrostatic speaker: A special type of speaker designed to avoid the problems inherent in cone speakers. They use a graphite-coated plastic membrane suspended between two perforated metal sheets. A high voltage of several thousand volts is applied to the membrane. The input signal is also raised to a high voltage by a transformer, and is applied to the perforated metal sheets.

encapsulation: A communications process in which a packet or frame is enclosed within another packet for the purpose of hiding the header of the encapsulated packet during transit through a network. At the destination, the receiving node recognizes the encapsulation header and recovers the original enclosed packet.

Ethernet: A popular local area network technology that is standardized by the IEEE as the 802.3 Carrier Sense Multiple Access with Collision Detection (CSMA/CD) standard. Ethernet was originally developed by two engineers at the Xerox Corporation and then developed further by a partnership of Xerox, DEC, and Intel. An Ethernet LAN typically uses coaxial cable, or special grades of UTP cable, as the transmission media.

European Broadcasting Union (EBU): A professional society that helps establish standards in the audio and broadcast industry in Europe, as well as other nations.

evaporator: A component of a heat pump or central air conditioning system located inside the home. It is also known as an *indoor coil* and functions as the heat transfer point for warming or cooling indoor air. It consists of a series of pipes connected to a furnace, or air handler, which blows the indoor air across the evaporator coil, causing the coil to absorb heat from the air. The cooled air is then delivered to the house through the ducting.

F

Fast Ethernet: The common name for the IEEE 802.3u Ethernet 100Base-T specification that transmits information at a wire speed of 100 Mbps.

fax: An abbreviation for facsimile machines used in home offices or businesses to send and receive documents using standard telephone lines, or dedicated digital lines. Faxing is a less expensive alternative to overnight package delivery, and can respond to situations in which a document is needed the same day for an urgent business transaction.

firewall: A system designed to prevent unauthorized access to, or from, a private network. Firewalls can be implemented in hardware or software, or a combination of both. Normally, a firewall is deployed between a trusted, protected private network and a public network. For example, the trusted network might be a corporate network and the public network might be the Internet.

FireWire: The product name for Apple Computer's version of the standard IEEE 1394 High Performance Serial Bus. It is used for connecting peripheral devices to personal computers. FireWire provides a single plug-and-socket connection on which up to 63 devices can be attached, with data transfer speeds of up to 400 Mbps.

fixed wireless access systems: Fixed wireless includes facilities for broadband service originating from ground-based transmission sites as well as satellites. Similar to satellite service, it is an alternative to technologies that use wired connections such as telephone and coaxial cable networks. The two FCC-licensed commercial services available in the fixed wireless service are called multichannel multipoint distribution service (MMDS), and local multipoint distribution service (LMDS). The multichannel multipoint distribution service can provide Internet-access downlinks over a distance of about 50 km from a central-transmitter site, using radio frequencies in the 2.1 and 2.6 GHz band. LMDS services use radio frequencies above 10 GHz—frequencies referred to as millimeter waves, at 28 GHz and 38 GHz. However, these are fiber-optic network replacement technologies intended for higher-density urban areas.

flat-panel display: A common name used to describe plasma or LCD displays. Flat-panel displays get their name from the type of construction that sets them apart from CRT displays. They are suitable for laptop computers and provide portable computing, with the technology to make them truly portable and lightweight. Flat screens are a special type of CRT display.

frequency: Within the context of wireless communications terminology, the number of cycles or reversals of an electromagnetic field that occur in one second. Frequency also expresses the quantity of units of information transmitted in a given time period, expressed in hertz.

front projection: A method of viewing video images that utilizes a video projector and a separate pull-down screen to show the projected image, similar to a movie screen.

G

gateway: A special home network interface device, similar to a router, that connects a single Internet access line such as DSL, cable, or telephone to the home computer network. The residential gateway serves as the communication hub for the entire suite of home networking devices deployed throughout the home. Gateways provide integrators with two functions — one is to connect the home network to the Internet, and the other is to act as a central communications interface resource for all in-home digital devices, ranging from lighting, security, and HVAC, to phones, home entertainment components, and personal digital assistants. Various types of residential gateways are available, based on the gradients of functionality required. Most provide capability for advanced data, voice, and video routing within the home. Most gateway products also include a firewall. The gateway is normally located near the distribution panel to simplify the connections from the gateway, to the home network patch panel, and the incoming Internet access port.

G.DMT (G.Discrete Multitone): A type of ADSL service defined in ITU Recommendation 992.1. The main difference between G.Lite and G.DMT is bandwidth. Sometimes called full-rate ADSL, the G.DMT variety can download data at up to 8 Mbps, and send data upstream at up to 1.5 Mbps

G.Lite: A type of Universal DSL described in ITU Recommendation G.992.2. Sometimes called *splitterless DSL*, it is a form of ADSL that does not require a splitter installation at the subscriber location, but at the expense of lower data rates. G.Lite can be used with a distributed splitter configuration, but with added complexity and limited benefit. It represents the most consumer-friendly version of DSL. G.Lite equipment and service costs less than other varieties, and it reportedly has a do-it-yourself installation. G.Lite supports a maximum of 1544 Kbps downstream, and 384 Kbps upstream.

Gigabit Ethernet: A term used to describe the IEEE 802.3z 1000Base-X Ethernet specification. Gigabit Ethernet, a transmission technology based on the Ethernet frame format and protocol used in LANs, provides a data rate of 1 billion bits per second (1 gigabit). Gigabit Ethernet is carried primarily on optical fiber cable (with very short distances possible on copper media).

GPM: Abbreviation for "gallons per minute."

Grade 1: The grade level described in ANSI/TIA/EIA 570A (Residential Telecommunications Cabling Standard) that provides a generic cabling system that meets the minimum requirements for telecommunication services. It specifies a minimum of one twisted-pair cable and associated connecting hardware (Category 3) and one coaxial cable (75-ohm) configured in a star topology. This grade provides for telephone, TV (digital or analog), satellite, CATV, and low-speed data services.

Grade 2: The grade level described in ANSI/TIA/EIA 570A (Residential Telecommunications Cabling Standard) in which each cabled location outlet requires two Category 5 UTP cables and two 75-ohm coaxial cables plus, as an option, optical fiber cabling configured in a star topology. This grade provides a broader range of residential services than Grade 1.

graphic equalizer: An audio unit that includes a set of filters, each with a fixed center frequency that cannot be changed. The only control available is the amount of boost or cut in each frequency band. This boost or cut is usually controlled with sliders. This interface is intuitive because the frequency response of the equalizer resembles the positions of the sliders themselves.

Ground Fault Circuit Interrupter (GFCI): A type of electrical outlet used to protect people from electrical shock due to faulty extension cords, appliances, or tools. A GFCI detects any minor imbalance in the current flowing in the hot and neutral conductors that might be caused by a ground fault in the connected appliance, and disconnects the outlet automatically.

H

heat pump: A special type of HVAC system that can heat a home during the winter and cool it during the summer. By definition a heat pump is any device that accepts heat at one or more temperatures, and rejects heat at a higher temperature.

High-Definition Television (HDTV): An advanced television broadcast standard adopted by the ATSC that provides higher resolution, a 16:9 aspect ratio, and 5.1 surround sound channels.

Home Audio/Video Interoperability (HAVi): A vendor-neutral audio/video standard aimed specifically at the home entertainment environment. HAVi allows different home entertainment and communication devices (such as VCRs, TVs, stereos, security systems, and video monitors) to be networked together and controlled from one primary device, such as a PC.

HomePNA (HPNA): The HomePNA abbreviation stands for Home Phone Network Alliance. It is a standard based on a process that enables voice and data transmissions to share the available bandwidth of the telephone cabling in a home, without mutual interference. It uses frequency division multiplexing (FDM) techniques to divide this bandwidth of frequencies into several communications channels. The acronyms for phone-line networking and power-line networking standards organizations can be confusing. The standard adopted for using the phone lines for networking is derived from the standards group that developed it (Home Phone Network Alliance). The published standard is often referenced as HPNA 2.0. (See also "HPLA.")

HomeRF: A wireless LAN home technology standard supported by the HomeRF Working Group. It is not compatible or interoperable with the IEEE 802.11 family of wireless LANs. The founding members of the HomeRF consortium include, among others, Microsoft, Intel, HP, Motorola, and Compaq.

home-run wiring: A method used in modern automated home design in which all the wiring is routed from a central point in the house, usually in a wiring closet. At the central point, the wires terminate at various types of patch panels or cross-connect terminals. This point becomes the interface between home wiring and external telecommunications wiring. Home-run wiring is the basic design concept used in structured wiring.

horizontal resolution: Chrominance and luminance resolution (detail) expressed horizontally across a picture tube. This is usually expressed as a number of black-to-white transitions, or lines, that can be differentiated. Horizontal resolution is limited by the bandwidth of the video signal, or equipment.

HPLA: An acronym for a standard called the HomePlug Powerline Alliance. The name is derived from a group of companies that developed the standard for power-line networking. The standard is referenced as HomePlug 1.0 or sometimes HPLA 1.0.

Hybrid Fiber Coaxial (HFC): A network that includes both fiber-optic cable and coaxial cable as the transmission media. HFC networks are used by the cable TV industry for the distribution of digital video channels and high-speed Internet access services.

Hypertext Markup Language (HTML): Authoring software used on the World Wide Web. HTML is basically ASCII text surrounded by HTML commands.

HVAC: Abbreviation for heating, ventilation, and air conditioning.

I

IEEE 802 standards: A series of standards approved by the IEEE, as follows:

- 802.1 (High-level Interface)
- 802.2 (Logical Link Interface)
- 802.3 (CSMA/CD)
- 802.4 (Token Bus)
- 802.5 (Token Ring)
- 802.6 (Metropolitan Area Networks)
- 802.7 (Broadband LANs)
- 802.8 (Fiber Optics)
- 802.9 (Integrated Voice Data)
- 802.10 (LAN Security)
- 802.11 (Wireless Networks)
- 802.12 (Demand Priority Access (100VG AnyLAN))
- 802.13 (Not Used)
- 802.14 (Cable TV-based Broadband)
- 802.15 (Wireless Personal Area Network)
- 802.16 (Broadband Packet)

IEEE-1394: Also known as FireWire, IEEE-1394 is a very fast external serial bus standard that supports data transfer rates of up to 400 Mbps. Products supporting the IEEE-1394 standard have different names, depending on the company. Apple, which originally developed the technology, uses the trademarked name FireWire. Other companies use other names, such as i.Link (Sony) and Lynx (TI), to describe their IEEE-1394 standard products. A single 1394 port can connect up to 63 external devices.

Insulation Displacement Connection (IDC): The recommended method of telecommunications copper wire termination recognized by the ANSI/TIA/EIA 568A standard. These termination devices are the popular choice for home network star topology connection points. Commonly called punchdown connections, these connections require the use of a small punchdown tool to properly secure the cable to terminal block. Punchdown connections remove or displace the conductor's insulation as it is seated in the connector. During termination, the cable is pressed between two edges of a metal clip, which displaces the insulation and exposes the copper conductor. This ensures a solid connection between the copper conductor and the terminating clip.

Integrated Services Digital Network (ISDN): A type of broadband, all-digital access service offered by local exchange carriers. ISDN provides circuit-switched access to the public network for voice, data, and video transmission. Basic rate interface (BRI) and primary rate interface (PRI) are two types of ISDN service.

interlaced scanning: A type of scan used in television broadcasting in which half the screen image is developed on successive frames of odd and even lines.

Intermodulation Distortion (IM): Intermodulation distortion is a measured performance feature of an amplifier that is listed in the specifications. It is caused when two or more frequencies become mixed in a nonlinear device. New frequencies that are not part of the source information are generated and produce undesirable effects in the amplifier output.

Internet: A worldwide collection of networks that make up an international information source, where almost any client can access any server. The Internet was initially a U.S. government network called the ARPAnet, which was named from the developing organization, the Advanced Research Projects Agency.

Internet Assigned Number Authority (IANA): An organization that previously performed services for IP address assignments and related engineering services, which are now performed by ICANN.

Internet Corporation for Assigned Names and Numbers (ICANN): A nonprofit corporation that was formed to assume responsibility for the IP address space allocation, protocol parameter assignment, domain name system management, and root server system management functions previously performed under U.S. government contract by IANA, and other entities.

Internet Engineering Task Force (IETF): An international community of network designers, operators, vendors, and researchers concerned with the evolution of the Internet architecture, and the operation of the Internet. The actual technical work of the IETF is done in its working groups.

Internet Protocol (IP): A primary protocol included in the TCP/IP suite used for encapsulating other protocols, such as TCP and UDP. The IP header includes the source and destination addresses of the packet, and other routing information.

Internet Service Provider (ISP): A company that provides individuals, and other companies, access to the Internet and other related services such as Web site building, and virtual hosting. An ISP has the equipment and the telecommunication line access required to have a point-of-presence on the Internet for the geographic area served.

ionization detector: A type of sensor used in home smoke detectors. It functions by detecting the disturbance caused by smoke in a small ionization chamber.

IrDA: IrDA stands for Infrared Data Association. It is an international organization that creates and promotes interoperable infrared data interconnection standards for wireless networks. The name is associated with products that conform to the interconnection standard.

ISO: A name used to identify the International Organization for Standardization. Because "International Organization for Standardization" would have different abbreviations in different languages ("IOS" in English, "OIN" in French for Organisation Internationale de Normalisation), the organization decided to use a word derived from the Greek "isos," meaning "equal." Therefore, whatever the country, whatever the language, the short form of the organization's name is always ISO. It is often used incorrectly as an acronym for International Standards Organization.

J

Jini: (pronounced GEE-nee) A Java-based standard from Sun Microsystems, which enables a plug-and-play network based on TCP/IP. Jini makes home networks dynamic by allowing devices and services to enter and leave the home network without complicated setup procedures. Jini shields the user from the installation, configuration, and setup process. Because Jini technology is also platform-independent, devices are no longer limited by specific brands of software, processors, device drivers, or traditional networking protocols. Jini is a direct competitor to HAVi, as it also defines a protocol for a plug-and-play networks.

junction (electrical) box: A square, octagonal, or rectangular plastic or metal box that fastens to framing, and houses wires, receptacles, and switches.

K

key system: A type of privately owned telephone system used by small business firms and home offices. It is less expensive than a PBX, and is designed for organizations that need as few as 3 or 4 phones, to as many as 250 phones. Key system phones have buttons that select outside lines, and can call other phones internally.

Most key systems include a controller-type cabinet called a key service unit (KSU), and a set of proprietary telephones. The KSU holds all the system's switching components, the system power supply, outside lines, and internal station cards.

kilowatt-hour meter: A device installed by the local electrical power utility to measure the power consumed in a home. The meter records the cumulative power consumed in kilowatt-hours. A single-phase watt-hour meter is essentially an induction motor, whose speed is directly proportional to the voltage applied, and the amount of current flowing through it. The phase displacement of the current, as well as the magnitude of the current, are automatically taken into account by the meter. In other words, the power factor influences the meter's speed and the moving element (disk) rotates with a speed proportional to true power. The read-out dials are simply a means for counting the motor revolutions, and by proper gearing, are arranged to read directly in kilowatt-hours.

L

LAN: Acronym for local area network. Also used to describe wireless local area networks (WLANs).

laserdisc: An older analog video optical disc format, first introduced in 1974. Each side of a disc could hold about an hour of video, so feature films required a pause in the middle to flip the disc over. Some films are on two discs. The introduction of DVDs made laserdiscs obsolete for the video market.

Liquid Crystal Display (LCD): A type of flat display technology used in laptop computer design, calculators, PDAs, and the majority of flat-screen displays. LCD displays are available as active matrix, dual-scan, or passive matrix display types. An LCD uses the fact that certain organic molecules (liquid crystals) can be reoriented by an electric field. As these materials are optically active, their natural twisted structures can be used to alter the polarization of light on a flat screen.

LNB: A component used in satellite receiving antenna feed systems. LNB is an acronym for Low Noise Block-down converter (so called because it converts a whole band or "block" of frequencies to a lower band).

load: The maximum amount of electrical current (in amperes) required for all the electrical devices in a home. Load analysis is usually performed by an electrical contractor for new home or remodeling construction. Load is also used to determine the power consumed by household electrical appliances. The actual power used by the load is called *true* power, or just *power*, and is measured in watts. (Even though watts = volts x amps, apparent power is measured in VA to differentiate it from true power.)

Local Exchange Carrier (LEC): Any public telephone company in the United States that provides local telephone service. Some of the largest LECs are the Bell operating companies (BOCs), which were grouped into holding companies known collectively as the regional Bell operating companies (RBOCs), when the Bell System was broken up by a 1983 consent decree. LEC companies are also sometimes referred to as *telcos* or *telephone companies*. A *local exchange* is the local central office of an LEC. When you pick up a phone to make a call, the dial tone you hear is coming from the central office of your local exchange carrier. Telephone lines from homes and businesses terminate at a local exchange facility. Local exchanges connect to other local exchanges within a local access and transport area (LATA).

LonWorks: A multipurpose home networking protocol developed by the Echelon Corporation that can be implemented over any medium, including power lines, twisted-pair cable, wireless (RF), infrared (IR), coaxial cable, and fiber optics. It has no central controller but uses a system of intelligent nodes that communicate with each other. The LonWorks protocol was published as ANSI/TIA/EIA-709 in 1998.

low-voltage wiring: All residential cabling that is not associated with 240V/120V high-voltage alternating current (AC) primary power wiring used in all U.S. residential construction. Although no exact definition of low-voltage exists in residential cabling standards, it is generally accepted to include those circuits that are connected to low-energy current-limited sources, as described in Article 725 of the NEC. Low-voltage wiring includes all wiring for audio/video components, network data cabling, security sensors, phone lines, telecommunications systems, and low-voltage landscape lighting. This encompasses all residential wiring, with the exception of 120-volt AC wiring.

lux: The amount of light required to obtain a reasonable CCTV video camera image. One lux is approximately the light from one candle measured from one meter. Typical camera ratings range between 0.5 and 1.0 lux.

M

magnetic stripe card: A type of security card used to control access to restricted areas, or residential gated communities. It consists of a plastic card with a narrow strip of magnetic material fused to the back. Data stored on the strip as narrow bars form the basis of a binary code, which is used by the access control system reader.

MC cable: Cable with a flexible metal covering. MC cable is composed of THHN soft-drawn copper wire conductors, and an insulated grounding conductor. It is suitable for branch, feeder, and service power distribution in commercial and industrial applications, as well as in multifamily buildings, theaters, and other populated structures. The acronym *THHN* refers to heat-resistant thermoplastic (90° C) for dry locations. This is a reference to the type of insulation on the wiring, coded per the NEC.

middleware: Connectivity software with enabling services, that allows multiple processes running on home computers to interact across a network. The main requirements for middleware are independence of any specific networking technology, interoperability between devices of different manufacturers, common application programming interfaces (APIs) that allow application development and service provision by third parties, and a common user interface framework that supports the specific requirements of the residential users, and their environment.

motorized damper: A component of an HVAC system that regulates the amount of warm or cold air that enters a room, or an area.

Moving Picture Experts Group (MPEG): A set of video and audio compression standards. The MPEG format employs compression algorithms to remove redundant picture information from successive video scenes. The MPEG methodology compresses only key objects within a frame, every 15th frame. Between the key frames, only the information that changes from frame to frame is recorded. The MPEG standard includes compression specifications for both video and audio signals.

MMDS (Multichannel Multipoint Distribution Service): A type of fixed, wireless communication service that uses FCC-licensed radio spectrum. Although originally developed as a one-way wireless service to deliver television service as a competitor to cable TV companies, MMDS has been permitted, through relaxed federal telecommunications legislative action, to provide broadband Internet service to business and residential customers.

multiplexer: An electronic, high-speed switching device that combines information contained on several digital video, or data channels, into a single high-speed channel. Multiplexers are used with home surveillance systems for recording several channels simultaneously.

N

National Electrical Code (NEC): A document sponsored by the National Fire Protection Association (NFPA). The National Electrical Code covers the requirements for electric conductors, and equipment installed within, or on, public and private buildings, or other structures. This includes mobile homes and recreational vehicles; floating buildings; and other premises such as yards, carnivals, parking and other lots, and industrial substations. It also covers conductors that connect the installations to a supply of electricity, and other outside conductors and equipment on the premises, including optical fiber cable. It covers buildings used by electric utilities, such as office buildings; warehouses; garages; machine shops; and recreational buildings that are not an integral part of a generating plant, substation, or control center.

Most states require a permit and an inspection for electrical work. Even though following the NEC cannot guarantee safe electrical installations, it is the best guide available. However, every state might differ slightly in its requirements for inspection and code compliance.

National Electrical Contractors Association: An international building trades organization that includes over 70,000 electrical contracting firms. NECA has 120 U.S. chapters, in addition to others in countries around the world.

National Electrical Manufacturers Association (NEMA): An organization of manufacturers of electrical equipment, including, but not limited to, wiring devices, wire and cable, conduit, load centers, pressure wire connectors, circuit breakers, and fuses. NEMA is the voice of the electrical industry, and through it, standards for electrical products are formulated. Generally, these standards promote interchangeability between products of one manufacturer, and similar products made by another manufacturer. In some cases, standards relating to product performance are also formulated by NEMA, but these are the exception rather than the rule.

National Fire Protection Association (NFPA): A U.S. standards organization that develops, publishes, and disseminates timely consensus codes and standards intended to minimize the possibility and effects of fire and other hazards. Virtually every building, process, service, design, and installation in society is affected by NFPA documents.

National Television System Committee (NTSC): The organization that developed the U.S. broadcast television standards. NTSC has also become the name of this group of standards. Other countries, such as Canada, Mexico, and Japan, have also adopted the NTSC broadcast standard.

Network Address Translation (NAT): A software feature used by gateways and routers to interconnect a private network that uses unregistered IP addresses with a global IP network that uses a limited number of registered IP addresses. NAT hides private IP addresses from the Internet, while still permitting the computers on that network to access the Internet.

Network Interface Card (NIC): A network interface card is a printed circuit board, with a connector that plugs into a PC expansion slot. A NIC contains a transceiver for sending and receiving data frames on a network, as well as the Data-Link layer hardware needed to format the sending bits, and to decipher received frames.

NH cable: A type of cable used in electrical wiring that is halogen-free (NH stands for nonhalogen). The noncorrosive material chosen enables this cable to be used for fixed installations in public buildings, and in governmental installations, where halogen-free products are demanded. The jacket is made of flame-retardant polyolefin material.

NM cable: Abbreviation for nonmetallic (NM) cable, the most widely used electrical power cable for indoor wiring. Romex is a brand name for a type of plastic insulated wire. It is often called nonmetallic sheath; however, the formal code name is NM cable. This type of wire is suitable for stud walls, on the sides of joists, and other areas that are not subject to mechanical damage or excessive heat. Newer homes are wired almost exclusively with NM wire.

O

octet: Eight bits in an IP address. An octet is thus an eight-bit byte. Since a byte is not eight bits in all computer systems, *octet* is used as a precise reference to eight bits. Technically, an octet represents any 8-bit quantity. By definition, an octet ranges in mathematical value from zero to 255. Typically, an octet is also a byte, but the term "octet" came into existence because historically, some computer systems did not represent a byte as eight bits.

optoelectronic: Describes any device that functions as an electrical-to-optical or optical-to-electrical transducer, or an instrument that uses such a device in its operation.

oxidation reduction potential (ORP): A measure of a swimming pool water's overall capability to eliminate wastes. High oxidation is present when pollution is low, and the water is of high quality. ORP is rapidly becoming the standard means of testing and regulating pool water sanitation.

P

packet: A unit of information formatted for transmission over a network. A packet contains control and header information that corresponds to the OSI Network layer, or layer 3. *Packet* is often used interchangeably (and incorrectly) with *frame*.

passive infrared (PIR): A type of motion detector sensor used in home security systems. The term *infrared* comes from the sensor's capability to see areas of the optical spectrum outside the range of light frequencies discernible by human vision. PIR motion detectors operate on the principle that heat, or infrared emissions, emanating from a human entering an area, will disturb an otherwise stable infrared background and trigger an alarm. Any area that does not change abruptly, over time, has a normal infrared background temperature. A PIR is referred to as passive because it does not release any energy of its own, but instead, causes an alarm when the surrounding infrared background environment changes abruptly.

peer network: A peer network (or peer-to-peer network), as the name suggests, is where all computers on a LAN are peers, and have equal status concerning sharing rights and capabilities. Any individual machine can share data or peripherals with any other machine on this type of network.

Personal Communications Service (PCS): A second-generation U.S. digital cellular telephone service established by the Federal Communications Commission. PCS services use the 1900 MHz band. All the licenses for the use of the PCS frequency bands by various service providers were awarded through public auctions for services in the major metropolitan areas. The FCC has also allocated spectrum for unlicensed PCS services in the 1910 MHz – 1930 MHz band. The unlicensed PCS service accommodates a wide range of services for small in-building areas, such as data networking within office buildings, wireless private branch exchanges (PBXs), personal digital assistants, laptop computers, portable facsimile machines, wireless replacements for portions of the wire-line telephone network, and other types of short-range communications.

Personal Video Recorder (PVR): A video recorder similar to a Video Cassette Recorder (VCR), except it uses a hard disk instead of a magnetic tape as the recording media. PVRs are designed to record live TV broadcast streams "on the fly" and allow the viewer to pause a program, take a break, and resume watching a program, even though the broadcast has continued past that point.

Phase Alternate Line (PAL): A color television broadcast standard used in Europe. Broadcasts in NTSC format can have problems with color hues varying in unwanted ways because of phase shifts in the color subcarrier. PAL fixed that by reversing the phase on every other line, hence the name. Because TV came later to Europe than to the United States, this superior format was adopted there instead of NTSC.

phone-line splitter: A device installed on a telephone line that is connected to both high-speed broadband digital service and analog voice devices, such as telephones and fax machines. The splitter routes the broadband and analog signals on the telephone line to the correct device. Most splitters must be installed by the telephone company; however, some can be installed by the customer.

photoelectric switch: A special type of switch used for controlling lights, depending on the light intensity (lux value). This type of switch contains a special sensor that operates when the light intensity, or darkness, reaches a certain value. Photoelectric switches are typically used outdoors for landscape and security lighting. Most photoelectric switches are integrated in a single fixture that includes the sensor and lamp, and have adjustments to turn off lights after a number of hours, where it is unnecessary to have lighting from dusk to dawn.

Plain Old Telephone Service (POTS): A name often used to describe the analog public switched telephone network.

Plasma Display Panel (PDP): A type of advanced large-screen display technology that uses hundreds of thousands of tiny cells (pixels) containing minute amounts of an inert gas, sandwiched between two sheets of glass. Electrodes are placed in pairs on the inner side of the front plate. When electrically charged, they produce an ultraviolet beam, which activates the phosphorous coating of the cell, and transmits light through the glass surface. The brightness of the emission depends on the current passing through the ionized gas. Large and brilliant color images can be produced with different colors of phosphors. Plasma displays provide a totally flat screen design with excellent off-center viewing capability, and superior resolution.

plenum: A compartment or chamber to which one or more air ducts are connected, and that forms part of the air distribution system.

Point-to-Point Tunneling Protocol (PPTP): A network protocol that encapsulates Point-to-Point Protocol (PPP) packets into IP datagrams for transmission over the Internet, or other public TCP/IP-based networks. PPTP can also be used in private intranets.

polyethylene: A plastic used for manufacturing irrigation tubing.

pop-up sprinkler: A popular type of sprinkler head used for home lawn and garden irrigation systems. It is installed below the ground, and the sprinkler head remains out of sight when it is inactive. When the sprinkler system is turned on, a small portion of the head emerges above the surface to disperse water to the irrigation area. Spray head sprinkler pop-ups are designed to spray a small fixed area, whereas rotor types are normally used for large, unobstructed areas.

Power Line Carrier (PLC): The capability for the power-line wiring used in the electrical system of a home to carry a command signal for controlling lighting and appliances. The X-10 control protocol that uses PLC technology was invented to exploit this capability.

Pressure Vacuum Breaker (PVB): A type of backflow preventer used with irrigation systems.

Private Branch Exchange (PBX): A telephone switch owned by a private organization or home office user. The system is usually located at the owner's facility. The PBX provides phone services including internal calling and access to the public switched telephone network. It allows a small number of outside lines to be shared among all the people of the organization. Advanced PBX phone switches sometimes provide auto-attendant, voice-mail, and ACD (automatic call distribution) services for the organization.

programmable thermostat: A special type of thermostat that enables temperature adjustments to be scheduled according to the day of the week and the time of day. It controls both heating and air conditioning equipment with a single integrated control processor.

progressive scan: A high-definition TV (HDTV) format. The progressive scan system scans the total number of lines, 60 times per second. This means that you see the complete image displayed on your TV screen twice as often than in the interlaced scan method. This results in smoother motion in moving images, fewer motion artifacts, and no visible flicker. A progressive scan system with 720 lines of resolution is written as 720p. Progressive scan has been used for many years for computer display monitors. It is also employed for DVD players that play back DVD movies in the progressive scan mode, for viewing on advanced digital television receivers.

protocol: A formal set of rules that establishes the method of handling data transmissions in both wired, and wireless, networks.

psi: Abbreviation for "pounds-per-square-inch."

Public Safety Answering Point (PSAP): A location where emergency response dispatchers answer 911 calls. The dispatch operator at the PSAP obtains as much information as possible from the caller, and sends the appropriate help to the caller's location. The caller might be in distress and unable to provide location information. The caller's phone number, however, is available to the dispatch operator through a system called automatic number identification (ANI). ANI provides the receiver of a telephone call with the number of the calling phone. The method of providing this information is determined by the local exchange carrier.

pull station: A device, typically mounted on a wall, that is used to manually activate a fire alarm system.

punchdown tool: A tool used to connect wires to an insulation displacement connection (IDC) block. This type of termination is the recommended method of copper termination recognized by the ANSI/TIA/EIA 568A standard for UTP cable terminations. These termination devices are the popular choice for making home network star topology connection points. Commonly called punchdown connections, these connections require the use of a small punchdown tool to properly secure the cable to the terminal block. Punchdown connections remove or displace the conductor's insulation as it is seated in the connector. During termination, the tool presses the cable between two edges of a metal clip, which displaces the insulation and exposes the copper conductor. This ensures a solid connection between the copper conductor and the terminating clip.

R

Registered Jack (RJ): In the United States, telephone jacks are also known as *registered* jacks, sometimes described as RJ-XX. This series of telephone connection interfaces (receptacle and plug) is registered with the FCC. They originate from interfaces that were part of AT&T's Universal Service Order Code (USOC), and were adopted as part of FCC regulations (specifically Part 68, Subpart F, Section 68.502). The term *jack* sometimes means both receptacle and plug, and sometimes just the receptacle.

relay: An electrical switching device often used in low-voltage control circuits to switch higher voltages on and off, such as a motor or lighting circuit.

remote access: A type of telecommunications service that provides user access to a computer, or network, from a remote location. It's often used by employees on business travel, who need access to the corporate private network to read e-mail or to access shared files. Remote access is also a valuable tool for home network users who need to access automated home features from a remote location.

Remote Authentication Dial-In User Service (RADIUS): A client/server protocol that enables remote access servers to communicate with a central server on a LAN, to authenticate dial-in users and authorize their access to the requested system or service. RADIUS enables a company to maintain user profiles in a central database that all remote servers can share. It provides better security, allowing a company to set up a policy that can be applied at a single, administered network point.

reversing valve: A type of valve used by heat pumps to redirect the refrigerant flow. It enables the inside coil and outside coil to act as either an evaporator, or a condenser.

ribbon speaker: A special type of speaker where the input signal is applied to a foil ribbon suspended between magnets, or metal sheets. Ribbon speakers do not use a cone and voice coil; the input signal from the amplifier is applied to the foil ribbon. The varying electrical charge caused by the input signal forces it to be repelled from, or attracted to, the magnets, thereby moving air and producing sound. This design overcomes some of the deficiencies of cone speakers in the mid and upper ranges of the audio spectrum.

riser: A type of duct for holding wire and cable that connects one floor to another, usually in a commercial building, and penetrates fire-rated walls or floors. Cable installed in risers must meet specific fire standards published by the NFPA. This type of cable is referred to as riser cable.

RJ-11: A registered jack (RJ) connector used for terminating a telephone cable that connects the phone to a standard wall outlet. It is used by many home automation devices, and is the standard used for all residential phones lines.

RJ-31X connector: A type of registered jack used for attaching alarm systems to telephone lines. The connector takes control of the phone line during a security alarm condition.

RJ-45 connector: A type of registered jack that uses an eight-conductor connector, with a small locking pin. It's used to connect the network interface card in networked computers to a local area network.

rotor sprinkler: Rotor sprinkler heads are pop-up sprinklers that disperse water over a large circular area. Small rotors can cover radii of up to 50 feet, whereas large rotors cover radii up to 100 feet. The two types of rotor sprinklers are called impact, and gear-driven. They differ only in the mechanism used to move the head in a circular motion.

router: A component consisting of hardware and software that provides intelligent connections between networks. Routers operate at the Network layer (layer 3) of the OSI model, and are responsible for making decisions about which paths through a network the data packets will use.

S

S/PDIF: Sony/Philips digital interface format is a consumer version of the AES/EBU digital audio interconnection standard that is based on coaxial cable and RCA connectors.

S-Video: A video interface standard that is a compromise between component analog video and composite video, because it separates the luminance (brightness) and chrominance (color) information. It uses twin coaxial type cables integrated into a single cable and terminated with a DIN connector.

scene: The effect created by a lighting control system, when a number of lights dim and brighten to different intensities.

Screened Twisted Pair (ScTP) cabling: A new type of 100-ohm, twisted-pair cabling that evolved from the ANSI/TIA/EIA standard IS-729. ScTP cable types are being developed to meet new, more rigorous criteria for Category 6E and 7E cable.

SECAM (Sequential Electronique Couleur Avec Memoire): Translated as sequential color and memory, SECAM is a composite color television transmission system that potentially eliminates the need for both the color and hue controls. It has a 625-line vertical resolution and a frame rate of 25-frames-per-second. One of the color difference signals is transmitted on one line, and the other is transmitted on the second line. Memory is required to obtain both color difference signals for color. This standard is used in France, Eastern Europe, Africa, and Asia.

set points: Temperature settings used on residential programmable and communicating thermostats that indicate how high or low the user wants the temperature to be in a home.

Shared Wireless Access Protocol (SWAP): A protocol derived from the existing Digital Enhanced Cordless Telephone (DECT) standard. The SWAP specification describes wireless transmission devices and protocols for interconnecting computers, peripherals, and electronic appliances in a home wireless (HomeRF) network environment. The Shared Wireless Access Protocol referenced in the HomeRF standard should not be confused with the Wireless Applications Protocol (WAP), which is used primarily for wireless Internet access on cellular networks.

Shielded Twisted Pair (STP) cable: A type of network cable originally developed as an IBM cabling system. It was used to support Token-Ring LANs. The original specification for STP was released in 1984 and defined the 150-ohm STP cable types 1, 2, 6, 8, and 9 for support of frequencies up to 16 MHz. It also defined the 100-ohm Type 3 UTP cable, as well as Type 5 and 5J fiber-optic cables. Later, an enhanced IBM cabling system defined STP-A cable Types 1A, 2A, 6A, and 9A for support of a network standard called Fiber Distributed Data Interface (FDDI). The *A* suffix denotes the enhanced IBM cabling system. The original IBM cabling system was defined in IBM publication GA27-3773. The enhanced, or STP-A, cabling is defined in the ANSI/TIA/EIA 568A standard.

Simple Network Management Protocol (SNMP): A protocol used for network management and monitoring of network devices and their functions.

single-room system: A lighting control system designed specifically for the control of lights in one room, such as a home theater.

Small Office Home Office (SOHO): A generic term in the home networking industry that refers to products and services primarily intended for the small business community, and residential computer networks.

solenoid: An electromechanical device used with irrigation system control valves. A solenoid consists of an electromagnet that, when energized, moves a valve to the open position. A spring retainer moves the valve to the closed position when the voltage is removed from the solenoid.

spread spectrum: A digital radio transmission technology in which the bandwidth of the radio carrier signal is considerably greater than that normally required to convey the digital baseband information. Spread spectrum technology provides excellent resistance to interference from other radio signals, and intentional radio frequency jamming sources, at the expense of wider spectrum utilization.

spike: A short burst of energy superimposed on the AC power line, usually of very short duration. Magnitudes may range between 0.5 to 2000 volts, lasting for 8 to 10 milliseconds.

stand-alone: A term used to describe a home technology component's capability to operate independently of another component.

structured wiring: The term *structured cabling/wiring* refers to all the cabling and components installed in a logical, hierarchical way, with a central distribution box serving as a common termination point for all residential cabling. It's designed to be relatively independent of the computer or telecommunications network that uses it, so either can be updated with a minimum of rework to the home. Structured wiring uses a wiring/cable design in which each room in a residence is wired for dual coaxial and UTP cable outlets.

subwoofer: A type of speaker designed for use with surround sound multiple-speaker systems. These speakers are optimized to reproduce sound only in the lower frequencies of the audio spectrum region of 20 Hz to 80 Hz.

surround sound speaker system: A sound system that uses five speakers placed in special locations in the front and side listening areas, and a subwoofer to handle the low-frequency effects channel.

switcher: A video controller used with security systems to connect several surveillance cameras to a single monitor. The switcher can be programmed to cycle through all the cameras in a surveillance system, or to dwell on each camera for a specified length of time.

T

Telecommunications Industry Association (TIA): An organization accredited by ANSI to develop voluntary industry and home technology standards. The TIA's Standards and Technology Department is composed of five divisions, which sponsor more than 70 standards-setting groups.

telnet: A TCP/IP terminal emulation protocol that enables remote login (the capability to log in to another computer) from a local computer. Telnet is a common way to remotely control Web servers. Telnet is also a Unix command that starts the telnet program on a Unix computer. The telnet program is the software (based on the telnet protocol) that enables the user to log in to a remote computer, and use its resources. The most common reason for using a telnet program is to log in to a remote computer to use e-mail, or to run programs not available on the user's local computer. Remote login to a more powerful computer can provide access to a full range of Internet services and sophisticated programs that often cannot be run on a low-end personal computer. Telnet can also be used to access various library catalogs, and databases. Telnet programs (or clients) are available for a wide variety of computer systems.

three-way calling: A telephone subscriber service offered by local exchange carriers, long-distance carriers, and mobile wireless cellular carriers. It enables a subscriber to talk to two people, in two different locations, at the same time.

Three-way calling is initiated by calling the first party, pressing and releasing the switch hook to get a second dial tone, and dialing a second number. When the second party answers, the switch hook is pressed and released again. The three-way connection is then in place, and the call can proceed with all three parties connected.

three-way switch: A type of switch used to control a light fixture from two locations. An example of this type of installation is where a light needs to be controlled at the top and bottom of a stairway. A three-way switch is required at both locations.

tie wrap: A plastic tie for holding cables together or holding cable in place, usually in the interior of a wall, or in a cable tray.

Toslink: A popular consumer equipment fiber-optic interface, based on the Sony/Philips digital interface format (SPDIF) protocol, using an implementation first developed by Toshiba.

touchscreen: A type of technology that uses a cathode-ray tube (CRT), or a liquid crystal display (LCD) screen, as both a keypad and a display. Touchscreens are popular with kiosks, point-of-sale cash registers, library search terminals, and computer-aided design (CAD) and drafting applications.

Transmission Control Protocol/Internet Protocol (TCP/IP): The name applied to a suite of protocols developed by the Advanced Research Projects Agency (ARPA).

TRIAC (thyrister for AC applications): A type of solid-state device used in home lighting dimmer controls. TRIACs switch the AC voltage on and off, using a variable phase control principle, permitting a gradual increase or decrease in the percentage of the phase that is applied to the load.

true power: The actual power used by the load of an electrical circuit measured in watts. (Even though watts = volts x amps, apparent power is measured in VA to differentiate it from true power.) The relationship between true power and VA is represented by the cosine of the phase angle between the voltage and current. The relationship between real power (in watts) and apparent power (in VA) is expressed as power (watts) = cosine (phase) x apparent power (VA).

twisted pair: A communication medium consisting of two copper conductors insulated from one another, and twisted together to reduce the effects of electromagnetic interference. Twisted-pair cables comprise the majority of installed telephone and home LAN cables.

U

UF cable: A type of power cable designed for direct burial in the ground. It is often used for irrigation system valves and control devices. It has a typical operating voltage rating of 600 volts.

Uninterruptible Power Supply (UPS): The most widely used type of backup power system for servers, and other computers, in home networking. A UPS provides power to client computers and servers in a business location, or home office, for a brief period after a power failure to allow a gradual planned shutdown of computer systems. UPS specifications indicate the length of time as well as the amount of electrical power, the unit can deliver when commercial power is lost. The most important rating to be aware of is the volt-ampere (VA) rating, which is the product of the volts and amperes delivered to the load.

Universal Plug and Play (UPnP): A Microsoft standard. Using UPnP, a device can dynamically join a network, obtain an IP address, convey its capabilities, and learn about the presence and capabilities of other devices automatically. UPnP uses Internet and Web protocols to enable devices such as audio and video components, PCs, peripherals, UPnP appliances, and wireless devices to be plugged into a network and automatically know about each other. With UPnP, when a user plugs a device into the network, the device will configure itself.

Universal Serial Bus (USB): A computer standard for a serial interface designed to connect peripheral devices to personal computers. The USB Specification, Revision 2.0, covers three speeds: 480 Mbps, 12 Mbps, and 1.5 Mbps. The term *Hi-Speed USB* refers to just the 480 Mbps portion of the USB specification; the term *USB* refers to the 12 Mbps and 1.5 Mbps speeds.

Unshielded Twisted Pair (UTP) cable: A type of cable widely used in LANs. It has insulated conductors twisted together, leading to better electrical performance, lower crosstalk, and significantly higher bit rates than those available with untwisted cable. The grading system divides UTP cable into separate categories. The spacing of the twists in a given length, along with fire code and installation criteria, dictate the official category rating to which the cable belongs, and the bandwidth that can be supported.

V

Video Home System (VHS): A standard for video tape systems developed by JVC in 1976. With VHS processing, the intensity information (luminance) and color information (chrominance) are filtered separately to reduce the bandwidth, and are then remixed when played back. This filtering, and the fact that the playback format is composite video, serve to reduce the quality of the image.

Virtual Private Network (VPN): A secure, private communication connection that is used by two or more devices across a public network (like the Internet). These VPN devices can be either a computer running VPN software, or a special device like a VPN-enabled router.

Voice over IP (VoIP): A voice communications system that uses the Internet Protocol. VoIP systems send voice information in digital form as discrete packets, rather than in the traditional analog circuit-switched protocols of the public switched telephone network. VoIP is also a term used in IP telephony for a set of facilities managing the delivery of voice information using the Internet Protocol. A major advantage of VoIP, and Internet telephony, is that they avoid the tolls charged by ordinary telephone service.

W

Web pad: A wireless, portable, tablet-shaped, consumer-focused digital device usually combined with a touchscreen. It has a browser-based interface to simplify and enhance the home control experience. The two components of a Web pad are a portable LCD built into a tablet-shaped device, and a base station that is plugged in to the home network, which sends and receives wireless (RF) transmissions. Transmissions between the Web pad and the base station use RF, or wireless, home networking technologies, such as HomeRF, Bluetooth, or IEEE 802.11b, and consequently have a limited range of 150 feet.

Wide Area Network (WAN): A telecommunications network encompassing a large geographical area. Usually, WANs are considered to cover distances beyond a metropolitan area.

Wi-Fi: An abbreviation for Wireless Fidelity. Wireless networking equipment that meets the 802.11 standard is approved to use the Wi-Fi label. It is used generically when referring to any type of wireless 802.11 network device, whether 802.11b, 802.11a, dual-band, and so on. The term is promulgated by an organization called the Wi-Fi Alliance. Any products tested and approved as Wi-Fi Certified (a registered trademark) by the Wi-Fi Alliance are interoperable with each other, even if they are from different manufacturers. A user with a Wi-Fi Certified product can use any brand of access point with any other brand of client hardware that also is certified. Typically, however, any Wi-Fi product using the same radio frequency (for example, 2.4 GHz for 802.11b or 11g, and 5 GHz for 802.11a) will work with any other product, even if it's not Wi-Fi Certified.

World Wide Web Consortium (W3C): An organization created to promote interoperability and establish an open forum for discussion of the World Wide Web (WWW). In October 1994, Tim Berners-Lee, inventor of the Web, founded the World Wide Web Consortium (W3C) at the Massachusetts Institute of Technology Laboratory for Computer Science, in collaboration with the European Organisation for Nuclear Research (CERN), where the Web originated, with support from the Defense Research Projects Agency (DARPA) and the European Commission.

X

X-10: A powerline carrier (PLC) protocol that enables compatible devices throughout the home to communicate with each other via the existing 120 Vac wiring in the house. Using X-10, you can control lights and virtually any other electrical device from anywhere in the house, with no additional wiring. The protocol name is derived from *Experiment 10*, which was a successful communications protocol perfected by Pico Electronics of Scotland in the 1970s, and was originally intended to control audio turntables. The original X-10 patent expired in December 1997. X-10 is now an open standard, and many manufacturers are developing new and improved X-10 products.

XDSL: A term used generically to refer to any of the digital subscriber line variants, such as HDSL, SDSL, RADSL, ISDL, or VDSL.l.

Z

zone: A designated area established in an automated home design for programming lighting scenes, or audio source material from a remote source or control point.

zone bypass: A feature used with home security systems to temporarily disable interior sensor systems when the residents are present, but wish to secure the perimeter area.

Index